Springer Texts in Education

More information about this series at http://www.springer.com/series/13812

Andy Liu

S.M.A.R.T. Circle Projects

 Springer

Andy Liu
Professor Emeritus
Department of Mathematical and Statistical
  Sciences
University of Alberta
Edmonton, AB
Canada

ISSN 2366-7672   ISSN 2366-7680   (electronic)
Springer Texts in Education
ISBN 978-3-319-56810-2   ISBN 978-3-319-56811-9   (eBook)
DOI 10.1007/978-3-319-56811-9

Library of Congress Control Number: 2017946034

Printed on acid-free paper

This Springer imprint is published by Springer Nature
The registered company is Springer International Publishing AG
The registered company address is: Gewerbestrasse 11, 6330 Cham, Switzerland

This book is dedicated to the past members of the S. M. A. R. T. Circle, in particular to the *Magnificent Seven:*

Mark Rabenstein

Graham Denham

Jason Colwell

Byung-Kyu Chun

Robert Barrington Leigh

David Dong-Uk Rhee

Mariya Sardarli

Preface

A Brief History of the S.M.A.R.T.Circle

A most beneficial side effect of the collapse of the former Soviet Union in 1992 was the migration of the Mathematical Circles across the Atlantic to the United States. Mathematical Circles, originated in Hungary during the nineteenth century, are a glorious tradition in Eastern Europe. They are organizations which discover and nurture young mathematical talents through meaningful extra-curricular activities.

The process took a few years, leading to the formation in 1998 of the Berkeley Mathematical Circle. With the support of the Mathematical Sciences Research Institute, the movement has caught fire in the United States, culminating in the formation of a Special Interest Group in the Mathematical Association of America under the leadership of Tatiana Shubin of San Jose State University.

Unbeknown to this community, a Mathematical Circle had existed in North America almost two decades earlier. The ultimate inspiration was still of Soviet origin, but the migration took place across the Pacific, via the People's Republic of China in the form of their Youth Palaces. This was the S.M.A.R.T. Circle in Edmonton, Canada, founded in 1981. The acronym stood for Saturday Mathematical Activities, Recreations & Tutorials.

I was born in China during the over-time sudden-death period of the Second World War, but moved to Hong Kong at age six. Thus I had never attended any session of any Youth Palace. However, I followed reports of their activities, and this fueled my interest in mathematics. The first mathematics book I had was a Chinese translation of Boris Kordemski's *Moscow Puzzles*, which was on their recommended reading list. An English version is now an inexpensive Dover paperback. Later, I acquired Chinese translations of several wonderful books by Yakov Perelman. Dover has published his *Figures for Fun* in English.

I came to Canada at age twenty, and eventually got a tenure-track position at the University of Alberta in 1980. That fall, I was invited to a general meeting of the Edmonton Chapter of the Association for Bright Children. My comment was that their activities seemed heavily biased towards the Fine Arts. Having put my foot in my mouth, I was obliged to take some concrete action. The next spring, the S.M.A.R.T. Circle was born.

During the first year, the members ranged from Grade 3 to Grade 6, because of the clientele of the A.B.C. However, to do meaningful mathematical activities, I preferred the children to be a bit more mature. So the Grade level rose by one each year, until in 1985, the members ranged from Grade 7 to Grade 10. Many of them stayed throughout this period.

As we moved away from the normal age of the clientele of the A.B.C., the Circle practically became an independent operation. This also became necessary because in 1983, we received a grant of $1,500 from the University of Alberta, arranged by Vice-President Academic **Amy Zelmer**. With the money, I built up a Circle Library containing mathematical books, games and puzzles. This was the only funding the Circle had received in its thirty-two year history.

We met on the University of Alberta campus from 2:00 pm to 3:00 pm every Saturday in October, November, February and March. A second class-room adjacent to the meeting room was open from 1:30 pm to 3:30 pm as the Circle Library. **Adrian Ashley**, a former Circle member, was hired at $5 an hour to look after it. There was a comedy of error in that for a while, his salary came out of the Student Union cafeteria account! They soon put a stop to that, but never bothered to claim readjustment.

Because of the members' tender ages, most came with their parents, and some parents stayed in Circle Library during the session. Members also had half an hour before and half an hour after the session to browse through. Sometimes, some younger members' attention span wandered during the session, and they would drift to the Circle Library for a few minutes.

In 1986, the three-year period of the grant ran out. As I closed the account, I turned the Circle Library over to the Faculty of Education. Then I started building a replacement out of my own pocket. Meanwhile, the A.B.C. had acquired new headquarters in the form of a house, where the basement was set up as a classroom. The Circle was invited to move its operation there. As a result, I restarted the session for A.B.C. members from Grade 3 to Grade 6 again. This went from 1:00 pm to 2:00 pm while the existing session for the older children ran from 2:30 pm to 3:30 pm. We had quite a few sibling pairs. Sometimes, one was in class in the basement while the other waited upstairs and played with mathematical games and puzzles from the new Circle Library. Sometimes, they sat in the same session despite any disparity in age.

In 1991, this arrangement came to an end, and the Circle moved back to the university campus. Only the Grade 7 to Grade 10 session survived the move. The meeting time was once again from 2:00 pm to 3:00 pm. A section at the back of the classroom was reserved for the Circle Library.

In 1996, there was a reverse migration of the Circle movement back across the Pacific, to Taiwan. My friend Wen-Hsien Sun of Taipei started the Chiu Chang Mathematical Circle, initially based on my model and using much of the material I had accumulated over a decade and a half. Both Circles closed in 2012, though mine was reincarnated as the J.A.M.E.S. Circle, standing for Junior Alberta Mathematics for Eager Students. It is run by my former student **Ryan Morrill**.

The activities of the S.M.A.R.T. Circle may be loosely classified into the following overlapping categories:

1. **Mathematical Conversations**;

2. **Mathematical Competitions**;

3. **Mathematical Congregations**;

4. **Mathematical Celebrations**.

At the beginning, the Circle activities consist only of the first two. The last two did not emerge until the second half of our Circle's thirty-two year history. For a description of these activities, see the companion volume *The S.M.A.R.T. Circle — Overview.*

The Mathematics Conversations are the heart and soul of the Circle. There is a Fall Session and a Winter Session each academic year. The Fall Session runs in October and November while the Winter Session runs in February and March. We meet every Saturday during those months from two to three in the afternoon at the University of Alberta. Each session consists of a minicourse plus a number of investigation topics. The latter lead to projects by Circle members, either independently or in small groups.

Over the thirty-two-year history of the Circle, many of our student projects have been published in scientific and education journals. This is by far the most successful aspect of our Circle. The material in this book is based on these publications.

Andy Liu,
Edmonton, Alberta, 2017.

Acknowledgement

I am very excited that **Springer-Verlag**, an institution in mathematics publishing, agrees to publish this volume. I am most grateful to their staff for encouragement and support, in particular, to Jan Holland, Bernadette Ohmer and Anne Comment. The technical team of Suganya Manoharan at Scientific Publishing Services, Trichy, India, has made significant contributions to the layout of the book.

Table of Contents

Chapter One: Error-Correcting Codes

Section 1. From the Alt Code to the Hamming Code

In a war, those who start it are usually at a safe distance from the front. Orders are sent to the troops via binary messages, sequences of 0s and 1s which may be interpreted according to a code by both the transmitter and the receiver. For example, if we want the troops to "Do nothing", we send the message 0. If we want the troops to "Drop the bomb", we send the message 1. This process is called **encoding**.

When the transmitted message is received, it must be interpreted. This process is called **decoding**. Here, 0 is decoded as "Do nothing" and 1 as "Drop the bomb". However, electronic transmissions may contain errors, due to imperfect equipments or enemy sabotage. We will assume that an error consists of a single digit-reversal per transmission.

In other words, the worst that can happen is that one of the 0s may turn into a 1, or one of the 1s may turn into a 0, but neither can happen twice and both cannot happen together in a single transmission. Even this is serious enough. In our example, if a 1 turns into a 0, we may not wake up in time to find that we have lost World War II. On the other hand, if a 0 turns into a 1, we may have started World War III while still fighting World War II.

We would like to have some way of telling whether we can trust what we have received. So we modify our simple code as follows: 00 would mean "Do nothing" and 11 would mean "Drop the bomb". If an error occurs, we will receive either 10 or 01, and we will know something has happened to the message. This is a prototype of what is known as an **error-detecting code**.

When we learn that an error has occurred, the natural thing is to ask headquarters to retransmit the message. If errors occur only infrequently, this is tolerable. If they occur often enough, it is at least a nuisance. Moreover, the request for retransmission is also sent electronically, and errors can occur there too.

What we would like is a code which not only tells us when something has gone wrong, but tells us exactly what has gone wrong. This is asking for a lot, but as is often the case, when we believe that something can be done, there may just be a way to do so. We now modify our code again, with 000 meaning "Do nothing" and 111 meaning "Drop the bomb". If we receive 001, 010 or 100, we do nothing. If we receive 110, 101 or 011, we drop the bomb. This is a prototype of what is known as an **error-correcting code**.

© Springer International Publishing AG 2018
A. Liu, *S.M.A.R.T. Circle Projects*, Springer Texts
in Education, DOI 10.1007/978-3-319-56811-9_1

Note that it is unrealistic to increase the length of the message and continue to assume that there is at most one error per transmission. So let us fix the length of any message to 15 binary digits, with at most one digit-reversal per transmission.

In some sense, even this is unrealistic since if one digit-reversal can occur, there is no reason why a second one cannot. What must be emphasized is that our assumption amounts to a mathematical model which simulates reality, but is not reality itself. We can get away with it if the probability of a single digit-reversal is high enough to worry us, but the probability of multiple digit-reversals is low enough to lose any sleep over.

Of the 15 digits, we may view some of them as conveying the intended message while the remaining ones are for security measure. The **efficiency** of a code using this 15-digit transmitter is defined as $\frac{n}{15}$. where n is the number of digits used for the message.

How can we extend our earlier examples with short messages to 15-digit messages? In an error-detecting code, we need to distinguish between two scenarios, whether the message contains an error or not. A single binary digit would allow us to do so. Hence the efficiency of such a code can be as high as $\frac{14}{15}$. Obviously. we cannot have $\frac{15}{15}$ as we will have no protection at all.

How can we encode the intended message 10110100010111? Should we add a 0 or a 1 as the fifteen digit? Let us reexamine the simple example earlier. To the message 0, we add a 0, and to the message 1, we add a 1. Note that we are not copying the message, but arrange for the coded message to contain an even number of 1s. Since 10110100010111 has 8 1s, we add a 0 to yield the coded message 101101000101110. This code is called the **parity-check code**, and the added digit is called the parity-check digit.

Decoding is straight-forward. Simply count the number of 1s in the received message. If no digit-reversal has occurred, this number is even as agreed. If a single digit-reversal has occurred, whether a 0 turned into a 1 or vice versa, the number of 1s will become odd. Thus a received message 001100101010110 contains an error. We do not know the intended message as any of the 15 digits may be the one which has been reversed.

In an error-correcting code on the 15-digit transmitter, we need to distinguish between sixteen scenarios, whether the message contains an error, and if so, which of the 15 digits has been reversed. We need four binary digits to do so because $2^4. = 16$. Thus the efficiency of such a code cannot be higher than $\frac{11}{15}$.

However, there is nothing in our earlier example to suggest how we can encode an 11-digit message. There, we stretch the message 0 to 000 and the message 1 to 111. This time, we are copying the message, twice, so that there are altogether three copies of the intended message. This means that the efficiency of this code is only as high as $\frac{5}{15}$, with the message 10110 encoded as 101101011010110. This code is called the **triple-repetition code**. It is also called the **Alt code** after its inventor (see [2]).

Decoding is based on the simple idea of majority rule. With at most one digit-reversal among three copies of the intended message, at least two copies must be correct. If all three agree, there are no errors. If not, the copy in the minority can safely be discarded. Thus a received message 010010000101001 contains an error in the seventh digit, and the intended message is 01001.

In 1984, Mark Rabenstein was an eighth grade student in Edmonton and a member of the SMART Circle. He had the following complaint about the Alt code: "If at most one digit-reversal can occur, why do we have to have three copies of the message? Wouldn't two be enough?"

"If an error has occurred so that the two copies are different, how do you know which is the correct one without reference to a third copy?" I asked.

"Just tag a parity-check digit to one of the copies, and you can tell if that one is correct."

That was a brilliant observation. In our fifteen-digit transmitter, we can use seven digits for the intended messages, repeat it a second time, and tag a parity-check digit to the second copy. For instance, the intended message 0011001 is encoded as 001100100110011.

Decoding is straight-forward. Compare the two copies of the intended message. If they agree, we can accept it. If not, apply the parity-check to the second copy to decide which one we would accept. For instance, if 101010110111011 is received, we see that 1010101 and 1011101 are different. Applying the parity-check to 10111011, we have an even number of 1s. Thus the received message contains an error in the fourth digit, and the intended message is 1011101.

The efficiency of $\frac{7}{15}$ is a big improvement over the Alt code. The **Rabenstein code** was published in [10].

The next break-through came in 1997. Han-Shian Liu (no relation to the author), then a sixth grade student in Taipei, was a member of the Chiu Chang Mathematical Circle. His (error-free) email message to me contained a gem.

"Mark was really smart to think of using the parity-check digit in an error-correcting code. It works so beautifully. Then I wonder whether I can make even more use of it. After experimenting with the idea for a while, I drew a tic-tac-toe board. (See Figure 1.1.) Each of the nine boxes contained a message digit. Then I added six parity-check digits, one for each row and one for each column."

For example, if the intended message is 011101001, we use the digits A to J are used to convey it, as shown in the grid below on the left. The digits K to Q are chosen so that the number of 1s in each row and each column is even.

A		D		G		N			A		D		G		N
	0		1		0		1			0		1		0		1
B		E		H		P			B		E		H		P
	1		0		0		1			1		0		0		1
C		F		J		Q			C		F		J		Q
	1		1		1		1			0		1		1		1
K		L		M					K		L		M		
	0		0		1					0		0		1		

Figure 1.1

Suppose the received message is as shown in the grid above on the right. Then the parity-check fails for the third row and the first column. It follows that the single digit-reversal occur at their intersection, namely, the box containing the digit C.

Han-Shian's code uses parity-check in two dimensions, and has an efficiency of $\frac{9}{15}$. This **Liu code** was published in [8], a paper which also contains an improved version that reaches the efficiency of $\frac{10}{15}$. So we are one step away from a perfect code with efficiency of $\frac{11}{15}$.

To mount this final assault, we use a set-theoretic representation of the tic-tac-toe board. Label the columns with a, b and c from left to right, and the rows with d, e and f from top to bottom. Then each box containing a message digit is represented by a two-element subset of $\{a, b, c, d, e, f\}$ while each box containing a parity-check is represented by a one-element subset. These two groups are separated from each other by a vertical line. Each parity-check digit is chosen so that the total number of 1s under columns containing the element which represents it is even. For example, the message 101111000 is encoded as shown in the chart below.

A	B	C	D	E	F	G	H	J	K	L	M	N	P	Q
a	a	a							a					
			b	b	b					b				
						c	c	c			c			
d			d			d						d		
	e			e			e						e	
		f			f			f						f
1	0	1	1	1	1	0	0	0	1	1	0	1	0	0
A	B	C	D	E	F	G	H	J	K	L	M	N	P	Q

Suppose the message in the chart below has been received. To decode it, we count the total numbers of 1s under the columns containing the elements a, b, c, d, e and f. They are 0, 3, 2, 2, 3 and 2 respectively. Since parity-check fails for the elements b and e, the error occurs at the element under the subset $\{b, e\}$. Hence the intended message is 000101010.

A	B	C	D	E	F	G	H	J	K	L	M	N	P	Q
a	a	a							a					
			b	b	b					b				
						c	c	c			c			
d			d			d						d		
	e			e			e						e	
		f			f			f						f
0	0	0	1	1	1	0	1	0	0	0	1	1	1	1

It is now clear why there is still room for improvement. Han-Shian had only made use of some of the non-empty subsets but not all of them. If we cut the set down to $\{a, b, c, d\}$, there are exactly fifteen non-empty subsets, four of which are singletons that give rise to the parity-check digits. For example, the message 10111100011 is encoded as shown in the chart below.

1	2	3	4	5	6	7	8	9	10	11	12	13	14	15
a	a	a	a		a	a	a				a			
b	b	b		b	b			b	b			b		
c	c		c	c		c		c		c			c	
d		d	d	d			d		d	d				d
1	0	1	1	1	1	0	0	0	1	1	0	1	0	0

Decoding is by the same method as in the Liu code. Suppose the message in the chart below has been received. The total numbers of 1s under the columns containing the elements a, b, c and d are 3, 5, 4 and 6 respectively. Since parity-check fails for the elements a and b, the error occurs at the element under the subset $\{a, b\}$. Hence the intended message is 10111100011.

1	2	3	4	5	6	7	8	9	10	11	12	13	14	15
a	a	a	a		a	a	a				a			
b	b	b		b	b			b	b			b		
c	c		c	c		c		c		c			c	
d		d	d	d			d		d	d				d
1	0	1	1	1	0	0	0	0	1	1	0	1	0	0

This perfect code is due independently to Golay [3] and Hamming [4] and is commonly known just as the **Hamming code**. The description given here is from [1]. Since it was already known early in the history of error-correcting codes, the Rabenstein code and the Liu code are irrelevant from an application point of view. However, they are the work of students and have high pedagogical value. They form a sequence of plausible reasoning that could eventually lead a lesser mortal from the simplistic Alt code to the same discovery by the founding fathers of coding theory.

In summary, the first step is to replace three copies of the messages by two, and incorporating parity-check. The second step is to perform parity-check in two dimensions instead of one dimension. The third step is to represent the code in set-theoretic format instead of geometric format. The chronology is given in [5].

Section 2. Two Applications of the Hamming Code

We give two unexpected applications of the Hamming code (see [7]). Both of them have strong recreational flavor.

Alice and Michael, along with thirteen of their friends, enter in a team competition organized by a certain hi-tech company. They will be put respectively into rooms A to Q (there are no rooms I and O). Each room is considered to be in one of two states, 0 or 1, assigned completely at random. Once they are isolated in their rooms, the team members will be informed of the state of each room except their own. Simultaneously, each must either pass, or declare the state of her or his room. They will have no further communication with their teammates, and are not aware of the action taken by any of them. If everybody passes, the team will be disqualified. If at least one declaration is incorrect, the team will also be disqualified. On the other hand, if there is at least one declaration, and all declarations are correct, the team wins a prize.

Alice, Michael and friends are given a short time to come up with some strategy. For instance, they could designate Alice as the guesser and have everyone else pass. The probability of winning a prize would then be $\frac{1}{2}$. However, they would like to do better. Alice and Michael come up with the following strategy based on the Hamming code.

We first give an illustration. Suppose Alice is in Room G and Michael is in Room N, and the actual states are as shown in the chart below.

A	B	C	D	E	F	G	H	J	K	L	M	N	P	Q
a	a	a	a		a	a	a				a			
b	b	b		b	b			b	b			b		
c	c		c	c		c		c		c			c	
d		d	d	d			d		d	d				d
0	1	1	1	0	1	1	0	0	1	1	0	0	1	0

Alice is in Room G which corresponds to the subset $\{a, c\}$. So she applies the parity-check to the element b and to the element d. Both tests pass. Now she applies the parity-check to the element a and to the element c, without taking the information on her room, which is unknown to her. Both will pass if the state of Room G is 0. So Alice will declare the opposite state 1.

Michael is in Room N which corresponds to the subset $\{b\}$. So he applies the parity-check to the element a. It fails. So there is no need to apply the parity-check to the elements c and d. Michael just passes.

This illustrates how things work in general. A team member declares if, and only if, the state of her or his room can be chosen to correct the single error in the corresponding Hamming code, but the opposite state is declared.

If the original set-up contains no errors when treated as a Hamming code, every team member will make an incorrect declaration. If the original set-up contains an error, only the team member in the room corresponding to where the error occurs will declare, and the declaration will be correct.

Recall that in the Hamming code, the 4 protection digits are uniquely determined by the 11 message digits. Of the $2^4 = 16$ possible sequences for those 4 digits, only the one contains no errors. Hence $\frac{15}{16}$ of the time, the original set-up contains an error. It follows that the probability of winning a prize is $\frac{15}{16}$.

This application is given in [12] as the problem titled *Crowning the Minotaur*. A special case was presented in [6]. The next application, in which we continue the story of Alice and Michael, is based on a problem in the Fall Round of the 2007 International Mathematics Tournament of the Towns. Only the former source mentions the Hamming code.

To celebrate their success in the team competition, the fifteen friends have a party, during which Alice and Michael perform a magic trick. While Michael is out of the room, the audience chooses one of the fifteen letters from A to Q inclusive, but excluding I and O. Then the audience places fifteen coins in a row, arbitrarily deciding whether each should be heads or tails. Alice either leave them alone or turns over exactly one coin, and leaves the room while Michael is brought back in. By looking at the coins and without knowing which one Alice has turned over, Michael determines the letter chosen by the audience.

Let us give an illustration. We use 0 to stand for a coin which is heads and 1 for a coin which is tails. Suppose the audience chooses the letter K and places the coins as shown in the chart below.

A	B	C	D	E	F	G	H	J	K	L	M	N	P	Q
a	a	a	a		a	a	a				a			
b	b	b		b	b			b	b			b		
c	c		c	c		c		c		c			c	
d		d	d	d			d		d	d				d
1	0	1	1	1	0	0	0	0	1	1	0	1	0	0

Applying the parity-check, Alice finds that it fails for a and b but passes for c and d. Since the letters under column K are b and d, Alice wants the parity-check to fail for b and d but pass for a and c. So she needs to reverse the parity for a and d, This can be done by flipping the coin under the subset $\{a, d\}$. Hence she changes the 0 under column H to 1. Had the audience chosen the letter F, Alice would have left the coins alone.

When Michael returns, he applies the parity-check, and finds that it passes for a and c but fails for b and d. This tells him that the letter chosen by the audience is the one associated with $\{b, d\}$, namely K.

Section 3. Reed-Muller Code

We now take up the issue of the correction of multiple errors. Our primary example is the Reed-Muller code (see [9] and [11]), which may be considered as an extension of the Hamming code. For a fifteen-digit transmitter, it can correct up to three errors. We set up a chart as in the Hamming code, except that there is now an additional vertical line separating the two-element subsets from the others, as shown below.

Suppose the intended message is 00101. We will now add ten digits for protection. To see what digit we must add under the column $\{a, b\}$, we consider the digits under the other columns which contain $\{a, b\}$, namely, $\{a, b, c, d\}, \{a, b, c\}$ and $\{a, b, d\}$. Since they are 0, 0 and 1 respectively, and we want an even number of 1s, we add the digit 1 under the column $\{a, b\}$. The digits under the next five columns are chosen in an analogous manner, and the chart is them completed as in the Hamming code, as shown below.

a	a	a	a		a	a	a			a				
b	b	b		b	b			b	b		b			
c	c		c	c	c		c		c			c		
d		d	d	d		d		d	d				d	
0	0	1	0	1	1	0	1	1	0	1	1	0	1	0

Suppose the following transmission via the Reed-Muller code has been received, with up to three errors.

a	a	a	a		a	a	a			a				
b	b	b		b	b			b	b		b			
c	c		c	c	c		c		c			c		
d		d	d	d		d		d	d				d	
0	0	1	0	0	0	0	1	1	0	1	0	1	1	0

We now perform a parity-check on the subset $\{a, b\}$. The digits under the columns $\{a, b, c, d\}, \{a, b, c\}, \{a, b, d\}$ and $\{a, b\}$ are 0, 0, 1 and 0. Hence the test fails. Note that this is not saying that the digit under $\{a, b\}$ is an error. It says that either one of these four digits is an error, or three of the four are errors. Performing parity-checks on the other two-element subsets, we find that the test fails for $\{b, c\}, \{b, d\}$ and $\{c, d\}$ but passes for $\{a, c\}$ and $\{a, d\}$.

We now perform a parity-check on the subset $\{a\}$. The digits under the columns $\{a, b, c, d\}, \{a, b, c\}, \{a, b, d\}, \{a, c, d\}, \{a, b\}, \{a, c\}, \{a, d\}$ and $\{a\}$ are 0, 0, 1, 0, 0, 0, 1 and 0. Hence the test passes. Performing parity-checks on the remaining subsets, we find that the test fails for $\{b\}, \{c\}$ and $\{d\}$.

We use P, Q and R to denote the three possible subsets of $\{a, b, c, d\}$ which are errors. Then an odd number of them contain $\{a, b\}$, $\{b, c\}$, $\{b, d\}$, $\{c, d\}$, $\{b\}$, $\{c\}$ and $\{d\}$ while an even number of them contain $\{a, c\}$, $\{a, d\}$ and $\{a\}$.

So $\{a\}$ appears either 0 or 2 times in P, Q and R while each of $\{b\}$, $\{c\}$ and $\{d\}$ appears either 1 or 3 times. Because $\{a, b\}$ appears either 1 or 3 times and it cannot appear without $\{a\}$, $\{a\}$ must appear exactly 2 times, say in P and Q. Also, $\{b\}$ cannot appear 3 times as otherwise $\{a, b\}$ will appear twice, but it must appear together with $\{a\}$. We may assume that it appears only in P.

Since each of $\{b, c\}$ and $\{b, d\}$ appears 1 or 3 times, both $\{c\}$ and $\{d\}$ must appear in P. Since each of $\{a, c\}$ and $\{a, d\}$ appears 0 or 2 times, both $\{c\}$ and $\{d\}$ must appear in Q. Since $\{c, d\}$ appears 1 or 3 times, both $\{c\}$ and $\{d\}$ must appear in R also. This is consistent with each of them appearing 1 or 3 times. Hence the errors are $P = \{a, b, c, d\}$, $Q = \{a, c, d\}$ and $R = \{c, d\}$, and the correct message is 10110.

While the decoding procedure in our example seems rather *ad hoc*, it does have a firm theoretical basis (see [1]). Let us give a more mathematical analysis of the above example.

For any set S, we define S^2 to be the collections of all non-empty subsets of S of size up to 2. In the above example,

$$
\begin{aligned}
P^2 &= \{\{a\}, \{b\}, \{c\}, \{d\}, \{a, b\}, \{a, c\}, \{a, d\}, \{b, c\}, \{b, d\}, \{c, d\}\}, \\
Q^2 &= \{\{a\}, \{c\}, \{d\}, \{a, c\}, \{a, d\}, \{c, d\}\}, \\
R^2 &= \{\{c\}, \{d\}, \{c, d\}\}.
\end{aligned}
$$

The symmetric difference of a number of collections consists of all elements which belong to an odd number of these collections. The symbol for symmetric difference is Δ. In the above example,

$$
P^2 \Delta Q^2 \Delta R^2 = \{\{b\}, \{c\}, \{d\}, \{a, b\}, \{b, c\}, \{b, d\}, \{c, d\}\}.
$$

The subsets in this collection are precisely those for which the parity-check fails. Thus $P^2 \Delta Q^2 \Delta R^2$ may be considered as the pattern of parity disturbance caused by the errors P, Q and R.

This pattern of parity disturbance may be caused by a different group of errors, namely $\{b, c, d\}$, $\{a, b\}$, $\{a\}$ and $\{b\}$, in that

$$
\begin{aligned}
&\{b, c, d\}^2 \Delta \{a, b\}^2 \Delta \{a\}^2 \Delta \{b\}^2 \\
={}& \{\{b\}, \{c\}, \{d\}, \{b, c\}, \{b, d\}, \{c, d\}\} \Delta \{\{a\}, \{b\}, \{a, b\}\} \Delta \{\{a\}\} \Delta \{\{b\}\} \\
={}& \{\{b\}, \{c\}, \{d\}, \{a, b\}, \{b, c\}, \{b, d\}, \{c, d\}\}.
\end{aligned}
$$

However, this does not invalidate the Reed-Muller Code as the second group contains four errors, more than the three we are allowed for a 15-digit transmitter. We claim that if two different groups of errors cause the same pattern of parity disturbance in a 15-digit transmitter, then one of them will consist of four or more sets.

Since they cause the same pattern of parity disturbance independently, they must cause no parity disturbance when acting together. After removing common sets from the two groups, we are left with a non-empty collection since the two groups of errors are different. We now prove that in a 15-digit transmitter, the number of sets in any non-empty collection which causes no parity disturbance is at least seven. It follows that four or more of them must come from the same group of errors, and our claim would be justified. We consider four cases.

Case 1. The collection consists only of 1-element sets.
Since there is at least one of them, the parity for the lone element in this set will be disturbed.

Case 2. The collection contains at least one 2-element set but no 3-element or 4-element sets.
The parity for the pair of elements in a 2-element set will be disturbed.

Case 3. The collection contains at least one 3-element set but not $\{a, b, c, d\}$. We may assume that the 3-element set is $\{a, b, c\}$. By itself, it will disturb the parity of $\{a, b\}$. To nullify this, the collection must contain either $\{a, b, d\}$ or $\{a, b\}$, but not both. Similarly, the collection must contain exactly one of $\{a, c, d\}$ and $\{a, c\}$, and exactly one of $\{b, c, d\}$ and $\{b, c\}$. Thus there are four sets in the collection with 2 or 3 elements. Collectively, they will disturb the parity of each of $\{a\}$, $\{b\}$ and $\{c\}$, and we need to include these three 1-element sets, bringing the total to seven sets.

Case 4. The collection contains $\{a, b, c, d\}$.
As in Case 3, the collection must contain either one or three of the set in each column of the chart below.

$$
\begin{array}{cccccc}
\{a, b, c\} & \{a, b, c\} & \{a, b, d\} & \{a, b, c\} & \{a, b, d\} & \{a, c, d\} \\
\{a, b, d\} & \{a, c, d\} & \{a, c, d\} & \{b, c, d\} & \{b, c, d\} & \{b, c, d\} \\
\{a, b\} & \{a, c\} & \{a, d\} & \{b, c\} & \{b, d\} & \{c, d\}
\end{array}
$$

We consider five subcases.
Subcase 4(a). There are no 3-element sets in the collection.
Then we must include all six 2-element sets, bringing the total to seven sets.
Subcase 4(b). There is only one 3-element set in the collection.
We may assume that it is $\{a, b, c\}$. Then we must include $\{a, d\}$, $\{b, d\}$ and $\{c, d\}$. This in turn forces the inclusion of $\{a\}$, $\{b\}$ and $\{c\}$, bringing the total to eight sets.

Subcase 4(c). There are exactly two 3-element sets in the collection.
We may assume that they are $\{a, b, c\}$ and $\{a, b, d\}$. Then we must include
$\{a, b\}$ and $\{c, d\}$. This in turn forces the inclusion of $\{c\}$ and $\{d\}$, bringing
the total to seven sets.
Subcase 4(d). There are exactly three 3-element sets in the collection.
We may assume that they are $\{a, b, c\}$, $\{a, b, d\}$ and $\{a, c, d\}$. Then we must
include $\{a, b\}$. $\{a, c\}$ and $\{a, d\}$, already bringing the total to seven sets.
Subcase 4(e). All four 3-element sets are in the collection.
Then we must also include all six 2-element sets, already bringing the total
to eleven sets.

We give some additional examples on the Reed-Muller Code. As before,
we use P, Q and R to denote the three possible subsets of $\{a, b, c, d\}$ which
are errors.

Example 1.
Decode the received message sent under the Reed-Muller code.

a	a	a	a		a	a	a				a			
b	b	b		b	b			b	b			b		
c	c		c	c		c		c		c			c	
d		d	d	d			d		d	d				d
1	1	1	1	1	0	1	1	0	0	1	0	0	0	0

Solution:
The parity-check fails for $\{a, b\}$, $\{b, c\}$ and $\{b, d\}$. Thus each of $\{a\}$, $\{b\}$,
$\{c\}$ and $\{d\}$ appears an even number of times. Since all are featured in
$\{a, b\}$, $\{b, c\}$ or $\{b, d\}$, each appears exactly twice. Let $\{b\}$ appear in P and
Q. Each of $\{a\}$, $\{c\}$ and $\{d\}$ appears an odd number of times with $\{b\}$,
which means exactly once. Hence they must appear together in R. Since
each of $\{a, c\}$, $\{a, d\}$ and $\{c, d\}$ appears an even number of times, $\{a\}$, $\{c\}$
and $\{d\}$ must appear together again, this time with $\{b\}$. Hence the errors
are $P = \{a, b, c, d\}$, $Q = \{b\}$ and $R = \{a, c, d\}$, and the correct message is
01101.

Example 2.
Decode the received message sent under the Reed-Muller code.

a	a	a	a		a	a	a				a			
b	b	b		b	b			b	b			b		
c	c		c	c		c		c		c			c	
d		d	d	d			d		d	d				d
1	0	1	0	1	1	0	1	1	0	1	1	0	1	0

Solution:
All ten parity-checks fail. Hence $P = \{a, b, c, d\}$ and $Q = R = \emptyset$, and the correct message is 00101.

Example 3.
Decode the received message sent under the Reed-Muller code.

a	a	a	a		a	a	a				a			
b	b	b		b	b			b	b			b		
c	c		c	c		c		c		c			c	
d		d	d	d			d		d	d				d
1	1	0	0	1	1	0	1	1	1	0	0	1	1	0

Solution:
The parity-check fails for $\{b\}$, $\{c\}$, $\{a, b\}$ and $\{b, d\}$. Thus each of $\{a\}$, and $\{d\}$ appears an even number of times. Since the parity-check for $\{a, d\}$ passes, they must appear together twice, say in P and Q. Now $\{b\}$ must appear with $\{a\}$ an odd number of times and with $\{d\}$ an odd number of times. Hence it appears in exactly one of P and Q, sat P, and not in R. Since the parity-check for $\{b, c\}$ pass, $\{c\}$ does not appear together with $\{b\}$, and similarly, it does not appear together with $\{a\}$ or with $\{c\}$. Hence the errors are $P = \{a, b, d\}$, $Q = \{a, d\}$ and $R = \{c\}$, and the correct message is 11101.

Example 4.
Decode the received message sent under the Reed-Muller code.

a	a	a	a		a	a	a				a			
b	b	b		b	b			b	b			b		
c	c		c	c		c		c		c			c	
d		d	d	d			d		d	d				d
1	1	0	0	1	0	0	1	0	1	1	1	1	1	1

Solution:
The parity-check fails for $\{b\}$, $\{c\}$, $\{b, c\}$, $\{b, d\}$ and $\{c, d\}$. Thus each of $\{a\}$ and $\{d\}$ appears an even number of times. Since $\{a\}$ is not featured in $\{b, c\}$, $\{b, d\}$ or $\{c, d\}$, it does not appear at all. Since $\{d\}$ is featured in $\{b, d\}$ and $\{c, d\}$, it appears exactly twice, say in P and Q. Each of $\{b\}$ and $\{c\}$ appears once with $\{d\}$. Hence each appears exactly once. Since the parity check for $\{b, c\}$ fails, they appear together once, say in P. Hence the errors are $P = \{b, c, d\}$, $Q = \{d\}$ and $R = \emptyset$, and the correct message is 11000.

If in our chart listing the subsets of $\{a, b, c, d\}$, we put in a third vertical line separating all three-element subsets from the others (well, just $\{a, b, c, d\}$), we can correct up to seven errors. The message now consists of a single digit, and it is easy to see that encoding simply repeats it to yield a total of fifteen copies. We have come full circle and return to the majority rule (eight out of fifteen in this case) which is the basis for the Alt Code.

Exercises

1. Design an error-correcting code with efficiency $\frac{10}{15}$.

2. Random justice is applied to three prisoners. On Decision Day, each prisoner will be given a hat to wear, which may be black or white. He can see the other two hats but not his own. At some point, the Warden will call for a simultaneous declaration from each prisoner, which he must make without the benefit of knowing how the other two will declare. He must declare "pass", "black" or "white". If the declaration is indeed' the color of his hat, he is right. If it is the other color, he is wrong. The prisoners will only go free if at least one of them declares, and all those who declare are right. The three prisoners get together the night before Decision Day and discuss strategies. What can they do to make the probability of their going free as high as $\frac{3}{4}$?

3. For a 15-digit transmitter, the efficiency of the Reed-Muller Code is $\frac{5}{15}$. Can this be improved?

Bibliography

[1] N. Alon and A. Liu, An application of set theory to coding theory, *Math. Mag.* **62** (1989) 233–237.

[2] F. L. Alt, A Bell Telephone Laboratories' computing machine (I), *Math. Comput.* **3** (1948/49) 1–13.

[3] M. J. E, Golay, Notes on digital coding, *Proc. I.E.E.E.* **37** (1949) 657.

[4] R. W. Hamming, Error detecting and correcting codes, *Bell System Tech. J.* **29** (1950) 147–160.

[5] A. Liu, In Search of a Missing Link: A Case Study in Error-Correcting Codes, *Math. Mag.*, **32** (2001) 343–347.

[6] A. Liu, A Magic Trick with Eight Coins, *8th Gathering for Gardner Exchange Book*, Vol. 1 (2006) 131–133.

[7] A. Liu, Two Applications of a Hamming Code, *Coll. Math. J.* **40** (2009) 2–5.

[8] H.-S. Liu and A. Liu, Error-correcting codes (in Chinese), *Math. Media* **91** #3 (1999) 59–63.

[9] D. E. Muller, Application of Boolean algebra to switching circuit design and to error detection, *IEEE Trans. Comput.* **3** (1954) 6–12.

[10] M. Rabenstein, An example of an error correcting code, *Math. Mag.* **58** (1985) 225–226.

[11] I. S. Reed, A class of multiple-error-correcting codes and the decoding scheme, em IEEE Trans. Inf. Theory **4** (1954) 38–49.

[12] D. E. Shasha, *Puzzling Adventures*, W. W. Norton, New York, (2005) 35–37 and 160–163.

Chapter Two: Regular and Semi-Regular Polyhedra

Section 1. Regular Polyhedra.

A **polyhedron** is a three-dimensional figure bounded by a finite number of polygonal faces. Its literal meaning is a many-faced figure because *poly* means many and *hedron* means face. Thus a tetrahedron is a four-faced figure, which can only be the triangular pyramid. The plural form of polyhedron is polyhedra. Some human beings are bihedra.

We will assume that the polyhedra we deal with are *convex*. In such a polyhedron, the line segment joining any of its two points lies entirely in the polyhedron. Most of the polyhedra we encounter, such as prisms and pyramids, are in fact convex.

The *skeleton* of a polyhedron consists of its vertices and edges only, and it contains all the essential information about the polyhedron. Thus we will represent any polyhedron by its skeleton.

We can facilitate the drawing of the skeleton of a polyhedron by the following process. Imagine that the edges are made of elastic strings. Choose a face as the base and stretch its edges so that the projection of every other vertex onto this face lies within its interior. For example, the skeleton of the tetrahedron with base BCD and opposite vertex A can be drawn as shown in the Figure 2.1. Such a representation is called the **Schlegel** diagram of the polyhedron.

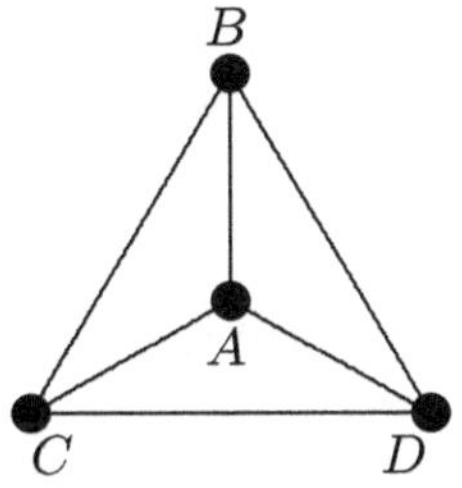

Figure 2.1

In his December, 1958 *Mathematical Games* column in *Scientific American* (see [2]), Martin Gardner wrote about the **Platonic** solids. They are named after the Greek philosopher Plato, and are the most pleasing of all polyhedra.

In a Platonic solid, all faces are the same kind of regular polygons and each vertex lies on the same number of faces. Thus there is perfect symmetry among the faces and among the vertices, both geometrically and combinatorially. It is not hard to see that there are only five Platonic solids. Suppose the faces are equilateral triangles. If we put three of them around each vertex, we have the regular tetrahedron as shown in Figure 2.1. If we put four of them around each vertex, we have the regular octahedron (double square pyramid) as shown in Figure 2.2.

© Springer International Publishing AG 2018
A. Liu, *S.M.A.R.T. Circle Projects*, Springer Texts
in Education, DOI 10.1007/978-3-319-56811-9_2

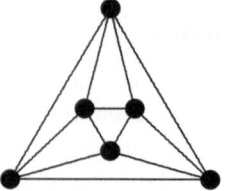

Figure 2.2

If we put five of them around each vertex, we have the regular icosahedron as shown in Figure 2.3. However, if we put six of them around a vertex, the configuration will be flat.

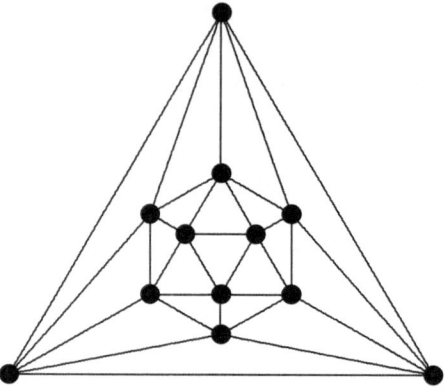

Figure 2.3

Suppose the faces are squares. If we put three of them around each vertex, we have the cube as shown in Figure 2.4. However, if we put four of them around a vertex, the configuration will be flat.

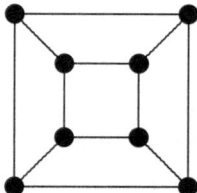

Figure 2.4

Suppose the faces are regular pentagons. If we put three of them around each vertex, we have the regular dodecahedron as shown in Figure 2.5. However, if we put four of them around a vertex, they will overlap.

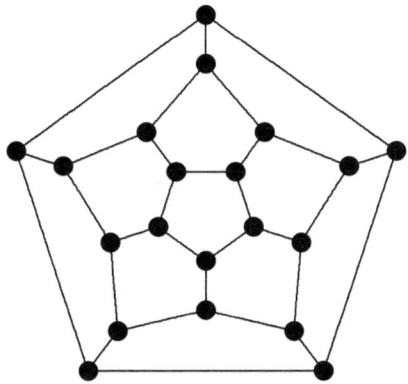

Figure 2.5

Suppose the faces are regular hexagons. Even if we place three of them around a vertex, the configuration will be flat. Thus there are indeed only five Platonic solids.

Schlegel diagrams are special cases of graphs. A **graph** is a collection of dots and lines, called **vertices** and **edges** respectively. Each edge connects two vertices. If these two vertices are identical, the edge is called a **loop**. If two edges connect the same two vertices, they are said to be **multiple edges**. The **degree** of a vertex is the number of edges which connects it to other vertices. Thus each edge contributes 2 to the total degree of a graph. A graph is said to be **connected** if any two vertices are accessible from each other via a sequence of edges.

If a graph can be drawn so that its edges meet only at the vertices, then it is called a **planar** graph. The Schlegel diagram of a polyhedron is also called a **polyhedral** graph, and must be planar. When drawn without crossing edges, a planar graph divides the plane unambiguously into regions. These regions are the **faces** of the graph. They correspond to the faces of the polyhedron. Clearly, non-planar graphs cannot be polyhedral graphs as the concept of a face is not well-defined. However, not all planar graphs are polyhedral graphs.

Note that in Figure 2.1, the base BCD is a face of the polyhedron, but seems to have disappeared as a face of the polyhedral graph. On the other hand, the graph has an infinite region which does not seem to be part of the polyhedron. To handle this apparent anomaly, imagine that the skeleton is drawn on the surface of a balloon, whose snout is contained in the face chosen as the base. If we stretch the balloon until the rim of its snout becomes a large circle enclosing a flat piece of rubber, we can see that the base has in fact become the infinite region.

Suppose a vertex of a polyhedron is surrounded in clockwise order by n faces, with x_i sides respectively for $1 \leq i \leq n$. We define its **vertex sequence** as $(x_1, x_2, \ldots, x_n)$. For a Platonic solid, all vertex sequences are identical, and it gives us a concise description of the solid. Thus the regular tetrahedron is (3,3,3), the regular octahedron is (3,3,3,3), the regular icosahedron is (3,3,3,3,3), the cube is (4,4,4) and the regular dodecahedron is (5,5,5).

Actually, a vertex sequence describes a whole class of polyhedra. For instance, (3,3,3) describes any tetrahedron, regular or otherwise.

A polyhedron is said to be **regular** if it satisfies the following two conditions:
(A) The sequences of all vertices are identical.
(B) All integers in the vertex sequence are identical.

How many regular polyhedra are there? There are at least five as all Platonic solids are regular polyhedra. Could there be other kinds? The geometric proof given earlier is no longer valid since we no longer require the faces to be regular polygons. Surprisingly, the answer is still five, but we need to give a combinatorial argument.

Our principal tool is a famous result due to the Swiss mathematician Leonhard Euler. He spent most of his active life in Prussia, under the patronage of Frederick the Great, and later in Russia, under the patronage of Catherine the Great. He had made contributions to many fields in mathematics, with an Euler's Formula in each of them. In particular, he was recognized as the father of graph theory.

Let V, E and F denote the numbers of vertices, edges and faces of a polyhedral graph. Then **Euler's Formula for Polyhedra** states that $V - E + F = 2$. It is also valid for connected planar graphs.

We shall prove Euler's Formula for Polyhedra in a slightly different form. A **component** of a graph is a connected subgraph which is not contained in any larger connected subgraph. In other words, each component is a connected piece of a graph. Denote by C the number of components. For connected graphs, we have $C = 1$.

We claim that for any planar graph, $V + F = E + C + 1$. Erase all the edges but retaining all the vertices. Initially, $V = C$, $F = 1$ and $E = 0$. We will reinstall the edges one at a time, so that E increases by 1 at each step. If the edge reinstalled in a particular step connects two vertices in different components, then C goes down by 1. If it connects two vertices in the same component, C remains unchanged but F goes up by 1 as an existing face is carved into two. Either way, the balance is maintained. At the end when all the edges have been reinstalled, we have $C = 1$ and $V + F = E + C + 1$ may be rewritten as $V - E + F = 2$.

As an application of Euler's Formula, we now prove that every planar graph without loops or multiple edges has a vertex of degree at least 5. Suppose to the contrary that every vertex has degree at least 6. Cut each edge into half-edges across its length. The total number of half-edges is exactly $2E$, and at least $6V$ by our assumption, so that $2E \geq 6V$. On the other hand, since there are no loops or multiple edges, each face is bounded by at least 3 edges, yielding $2E \geq 3F$. Substituting into Euler's Formula, we have $2 = V - E + F$
$le \frac{E}{3} - E + \frac{2E}{3} = 0$, and we have a contradiction.

Condition (A) in the definition of a regular polyhedra implies that all vertices of the polyhedron lies on the same number n of edges, and condition (B) implies that all faces of the polyhedron are bounded by the same number m of edges. Thus the vertex sequence of a regular polyhedra consists of n copies of m. Since every polyhedral graph has a vertex of degree at most 5, we see that $n \leq 5$. That $m \leq 5$ can be proved similarly. Thus we have nine cases, as shown in the chart below.

	$m = 3$	$m = 4$	$m = 5$
$n = 3$	Standard Tetrahedron	Cuboid	Standard Dodecahedron
$n = 4$	Standard Octahedron	Impossible	Cases
$n = 5$	Standard Icosahedron		

To prove that the four cases marked impossible are indeed so, we need a preliminary result: $nV = 2E = kF$. Cut each edge in halves at its midpoint. Each of the V vertices is attached to n half-edges, so that the total number of half-edges is nV. On the other hand, the total number of half-edges is clearly $2E$ since each of the E edges is cut in halves. It follows that $nV = 2E$. Similarly, we can prove that $2E = kF$ but cutting each edge in halves along its midline and count the half-edges in two ways as before. Recall that we have used this argument in proving that K_5 and $K_{3,3}$ are non-planar.

Substituting into Euler's Formula, we have $\frac{2}{n}E + \frac{2}{m}E - E = 2$ or

$$\frac{1}{n} + \frac{1}{m} = \frac{1}{2} + \frac{1}{E} > \frac{1}{2}.$$

If $n = m = 4$, we have only $\frac{1}{n} + \frac{1}{m} = \frac{1}{2}$. In the other three cases, we have $\frac{1}{n} + \frac{1}{m} < \frac{1}{2}$. All contradict the above inequality.

Section 2. A Polyhedral Metamorphosis.

Circle members participate in the International Mathematics Competition, a sample paper of which is given in [5]. An important event in the competition is the Cultural Evening, during which each country presents a short performance that highlights their heritage. As a multi-cultural nation, Canada have had a hard time finding suitable things to do. Finally, in 2010 when the competition was in Inchon, South Korea, we decided to express ourselves in a universal language, namely, mathematics.

The performance is described in [4]. Ten students use six strings to construct the skeleton of each of the five Platonic solids in a sequence of continual transformation. Start with four students forming the tetrahedron. At some point, six students join in. After a while, the original four drop out. Eventually, the remaining six students form the octahedron. During this sequence, all of the other Platonic solids appear. At the end, the original four take over from the final six and restore the tetrahedron! This is adapted from the design by Karl Schaffer [6].

Step 1. Construction of the Tetrahedron

We start of with four students identified as N(orth), S(outh), E(ast) and W(est). Each designates one hand as the U(pper) hand and the other hand as the L(ower) hand. N and S hold out their U hands while E and W hold out their L hands. String 1 is held between UN and LW, string 2 between UN and LE, string 3 between LW and LE, string 4 between LW and US, string 5 between LE and US, and string 6 between UN and US. The completed tetrahedron is shown in Figure 2.6, with string 6 drawn in such a way to facilitate the next step.

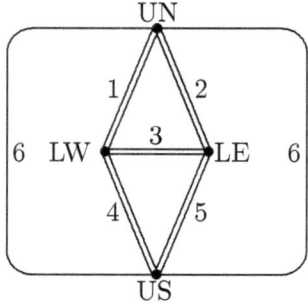

Figure 2.6

Step 2. Transformation into the Cube

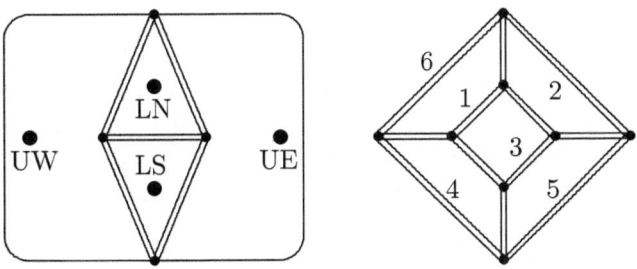

Figure 2.7

Each of the four students holds out the other hand and places it at the center of one of the four faces of the tetrahedron, as shown on the left side of Figure 2.7. Each of these hands will grab the three sides of the triangular face. The end result is a cube, as shown on the right side of Figure 2.7. Each string forms a face of the cube.

Step 3. Transformation into the Dodecahedron

We first redraw the cube as shown in Figure 2.8.

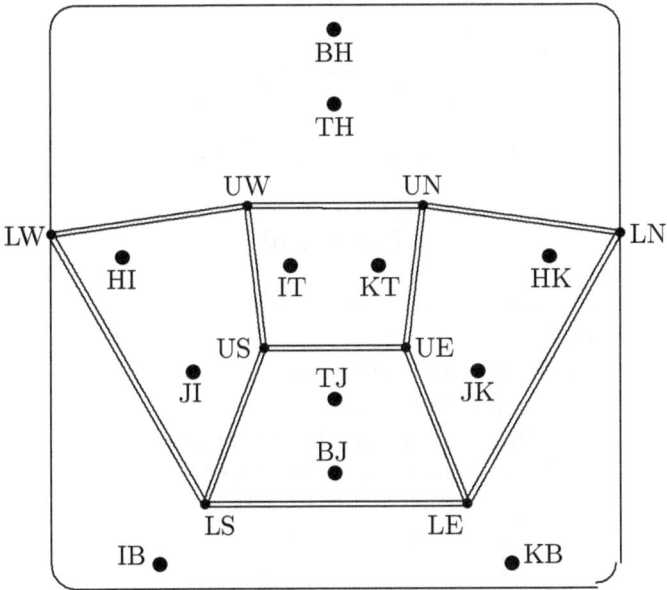

Figure 2.8

Now six other students enter the picture. They are identified as T(op face), B(ottom face), H (northwest face), I (southwest face), J (southeast face) and K (northeast face). Each of these students hold out both hands and places them symmetrically about the center of the assigned face of the cube. The line segment joining the two hands of each student is parallel to a side of the cube, and the segments on adjacent faces are perpendicular to each other. Each pair of these hands will grab the two sides of the square face parallel to the segment they form. Each hand will also grab the nearer one of the remaining two sides of the square face. The end result is a dodecahedron, as shown in Figure 2.9.

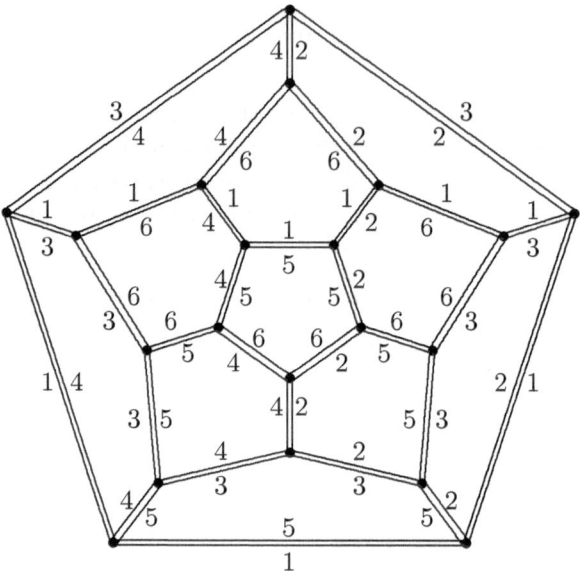

Figure 2.9

It should be emphasized that while each face of the cube is formed of one string, no part of this string is to be grabbed by the hands assigned to this face. Instead, the other four strings joining adjacent pairs of vertices of the face are grabbed, as illustrated in Figure 2.10. Failure to exercise the caution in the preceding paragraph will still produce a dodecahedron, but the whole structure will then fall apart in Step 4.

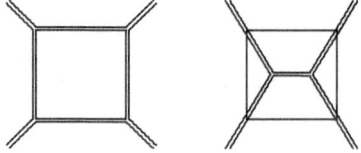

Figure 2.10

Step 4. Transformation into the Icosahedron

The original four students let go of their strings. The end result is an icosahedron, as shown in Figure 2.11.

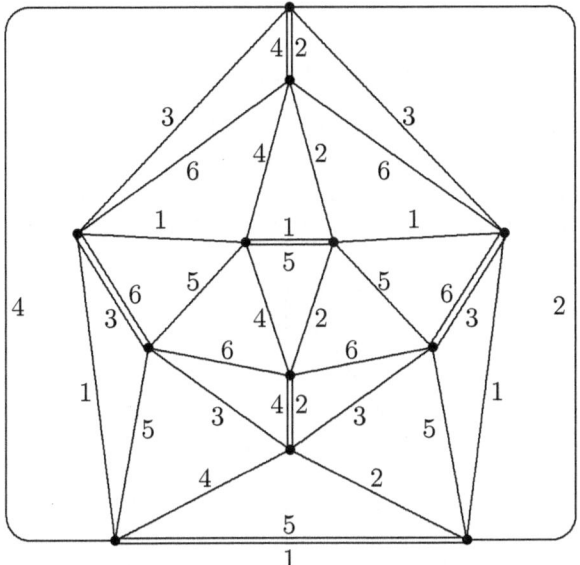

Figure 2.11

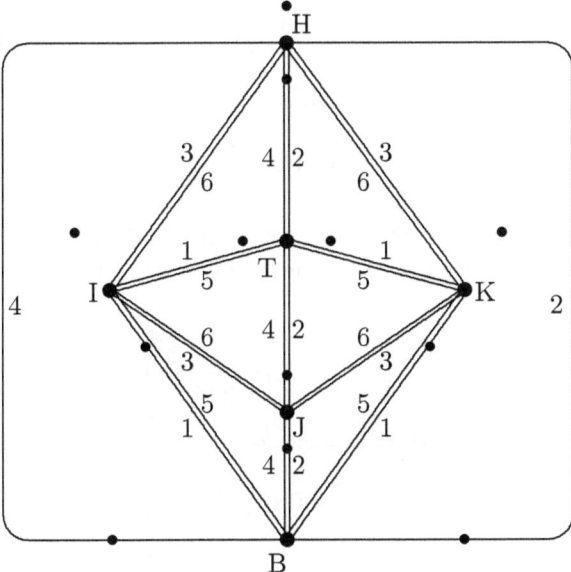

Figure 2.12

Step 5. Transformation into the Octahedron

Each of the remaining six students slides both hands together. The end result is an octahedron, as shown in Figure 2.12. Two strings which are opposite sides of the original tetrahedron now form the same square cross-section of the octahedron.

Step 6. Return to the Tetrahedron

The original four students N, S, E and W re-enter the picture. N puts the U hand in triangle HKT (north and top), S puts the U hand in triangle IJT (south and top), E puts the L hand in triangle JKB (bottom and east) and W puts the L hand in triangle HIB (bottom and west). This is shown in Figure 2.13.

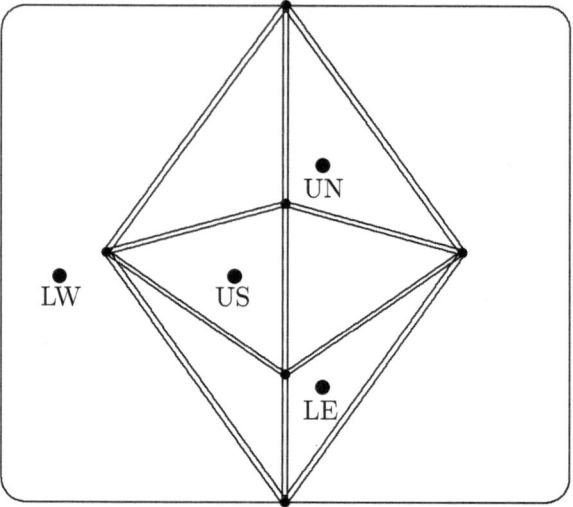

Figure 2.13

Each hands grabs the three strings it originally holds, and then the other six students let go of theirs. The end result is a tetrahedron, as shown in Figure 2.14.

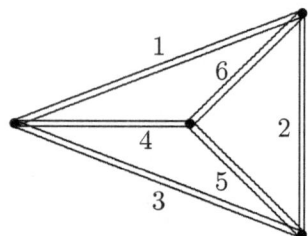

Figure 2.14

Section 3. Semi-Regular Polyhedra.

A polyhedron is said to be **semi-regular** if it satisfies condition (B) in the definition of a regular polyhedron. Clearly the Platonic solids are semi-regular, but they are now joined by infinitely many others. In Figure 2.15, we depict the case $n = 8$ from each of two infinite classes of semi-regular polyhedra — the prisms $(4, 4, n)$ and the antiprisms $(3, 3, 3, n)$, where $n \geq 3$. In particular, the cube is the order-4 prism, and the regular octahedron is the order-3 antiprism.

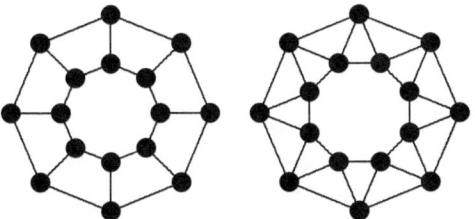

Figure 2.15

Apart from these three classes, there are other semi-regular polyhedra. Let us use a geometric approach to see if we can unearth some of them.

By slicing off, in a systematic manner, the corners of the regular tetrahedron, we obtain the truncated tetrahedron (3,6,6), as shown in Figure 2.16.

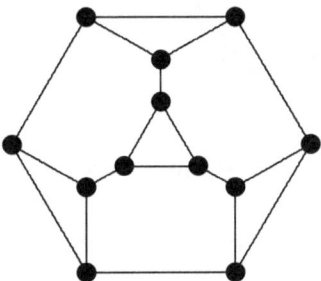

Figure 2.16

The same process applied to the cube, the regular dodecahedron, the regular octahedron and the regular icosahedron produces the truncated cube (3,8,8), the truncated dodecahedron (3,10,10), the truncated octahedron (4,6,6) and the truncated icosahedron (5,6,6), as shown in Figures 2.17, 2.18, 2.19 and 2.20.

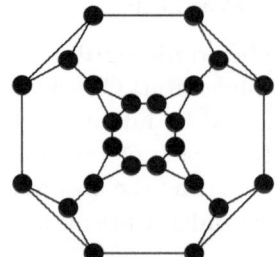

Figure 2.17

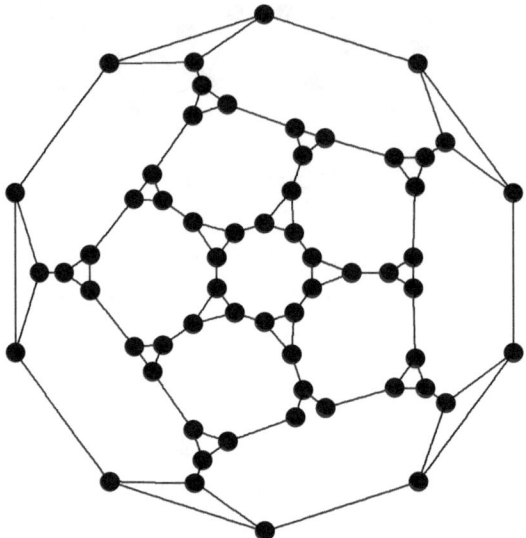

Figure 2.18

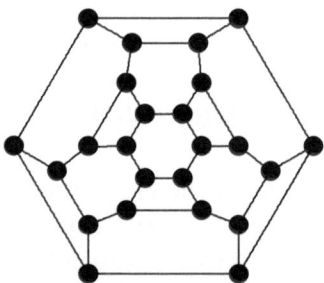

Figure 2.19

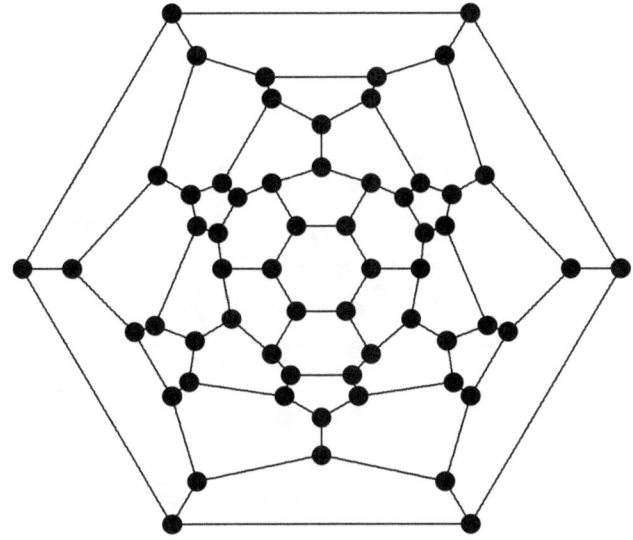

Figure 2.20

If we truncate the cube or the octahedron to the midpoints of the edges, we obtain the cuboctahedron (3,4,3,4). If we truncate the dodecahedron or the icosahedron to the midpoints of the edges, we obtain the icosadodeca-hedron (3,5,3,5). These are shown in Figures 2.21 and 2.22.

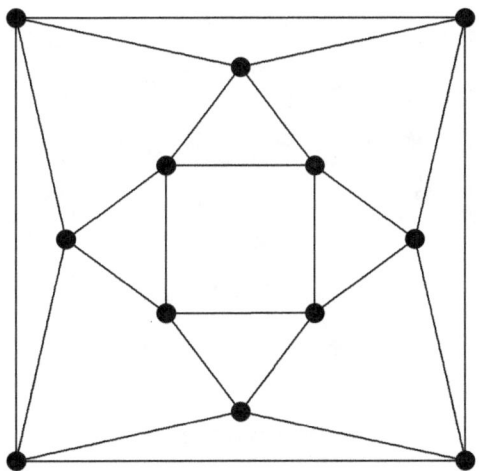

Figure 2.21

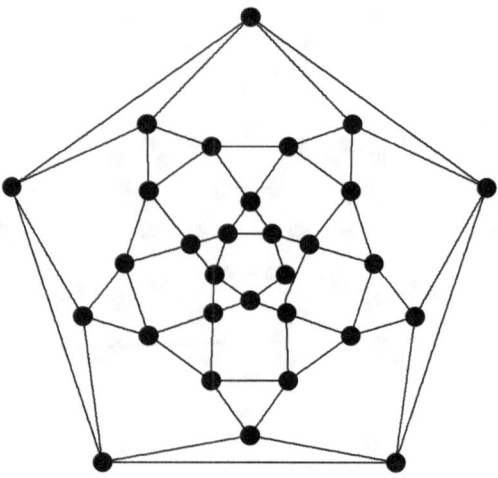

Figure 2.22

Figure 2.23 shows the great rhombicuboctahedron (4,6,8) which are obtained by truncation from the cuboctahedron.

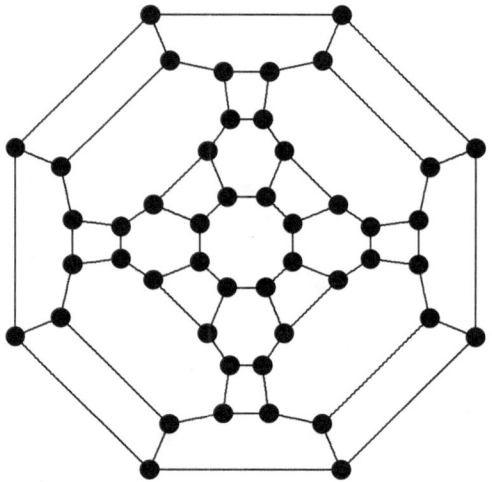

Figure 2.23

Figure 2.24 shows the great rhombicosadodecahedron (4,6,10) which is obtained by truncation from the icosadodecahedron.

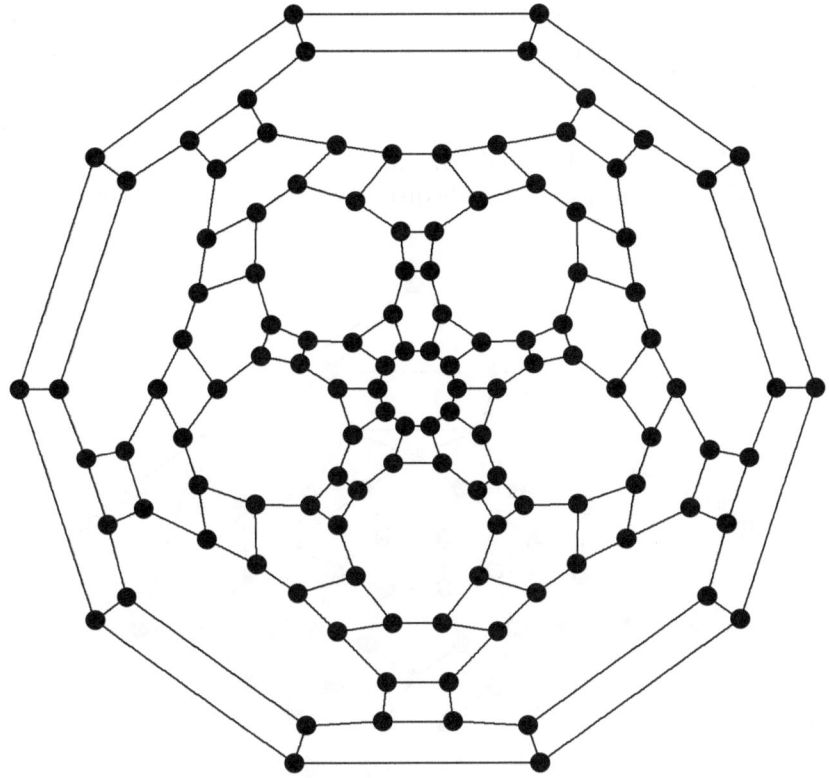

Figure 2.24

Figure 2.25 shows the small rhombicuboctahedron (3,4,4,4) which is obtained from the cuboctahedron by taking the truncation to the midpoints of the edges.

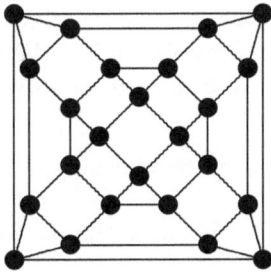

Figure 2.25

By rotating the octagon in Figure 2.25 45°, we obtain a polyhedron which has the same vertex sequence (3,4,4,4) as the small rhombicuboctahedron. It has much less geometric symmetry and is not considered to be a new semi-regular polyhedron. It is not obtained by truncation.

Figure 2.26 shows the small rhombicosadodecahedron (3,4,5,4) which is obtained from the icosadedecahedron by taking the truncation to the midpoints of the edges.

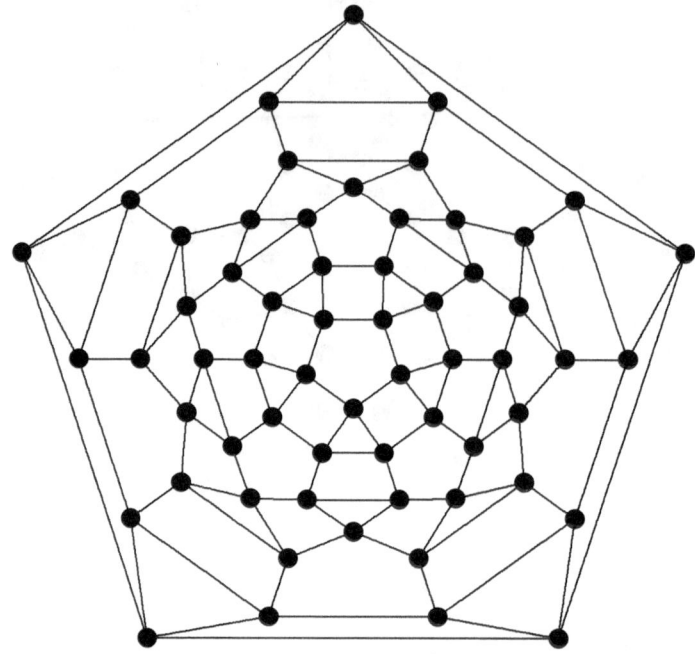

Figure 2.26

Are there some semi-regular polyhedra which are not obtained by truncation? In any case, how can we find all of them? To do so, we return to the combinatorial approach in Section 1 to prove our main result.

Theorem.
The vertex sequence of any semi-regular polyhedron is among the following:

(I) (3,3,3), (4,4,4), (5,5,5), (3,3,3,3), (3,3,3,3,3);

(II) $(4, 4, n), n \geq 3$;

(III) $(3, 3, 3, n), n \geq 3$;

(IV) (3,6,6), (3,8,8), (3,10,10), (4,6,6), (5,6,6), (4,6,8), (4,6,10), (3,4,3,4), (3,5,3,5), (3,4,4,4), (3,4,5,4), (3,3,3,3,4), (3,3,3,3,5).

The thirteen polyhedra in (IV) are called *sporadic* semi-regular polyhedra, and if their faces are regular polygons, they are called **Archimedean** solids, named after another Greek philosopher.

Note that the combinatorial approach yields two additional semi-regular polyhedra not obtained geometrically by truncation. They are the snub cube (3,3,3,3,4) and the snub dodecahedron (3,3,3,3,5). Both have two versions in opposite orientations, but they are not considered to be different semi-regular polyhedra.

To prove the Theorem, let there be t kinds of faces. For $1 \le i \le t$, let each face of the i-th kind be bounded by x_i edges, and let the number of such faces be F_i. Suppose the vertex sequence of the solid has length n and consists of λ_i copies of x_i for $1 \le i \le t$, with $\lambda_1 + \lambda_2 + \cdots + \lambda_t = n$.

Count the vertices of each face bounded by x_i edges. The total is $x_i F_i$. Each vertex has been counted λ_i times for a total of $\lambda_i V$. It follows that $F_i = \frac{\lambda_i}{x_i} V$. This is a generalization of our earlier result $nV = mF$. Moreover, we still have $nV = 2E$. Putting all these into Euler's Formula, we have $V + \frac{\lambda_1}{x_1} V + \cdots + \frac{\lambda_t}{x_t} V - \frac{n}{2} V = 2$ or

$$\frac{\lambda_1}{x_1} + \cdots + \frac{\lambda_t}{x_t} = \frac{2}{V} + \frac{n-2}{2}.$$

We call this the characteristic equation.

Each of n, x_1, $\ldots$, x_t is at least 3. We shall deduce from the characteristic equation that at least one of the x's is less than 6. Assuming the contrary, we have the following contradiction:

$$\frac{2}{V} + \frac{n-2}{2} = \frac{\lambda_1}{x_1} + \cdots + \frac{\lambda_t}{x_t} \le \frac{n}{6} \le \frac{n}{6} + \left(\frac{n}{3} - 1\right) = \frac{n-2}{2}.$$

Similarly, we can show that $n = 3$, 4 or 5. Assuming the contrary, we have the following contradiction:

$$\frac{2}{V} + \frac{n-2}{2} = \frac{\lambda_1}{x_1} + \cdots + \frac{\lambda_t}{x_t} \le \frac{n}{3} \le \frac{n}{3} + \left(\frac{n}{6} - 1\right) = \frac{n-2}{2},$$

We now divide the proof into three parts, for $n = 3$, $n = 4$ and $n = 5$ respectively.

Part One. $n = 3$.

Let the vertex sequence be (a, b, c) with $a \le b \le c$, where $3 \le a \le 5$. We consider four cases.

Case 1. $a = b = c$.

Since $3 \le a \le 5$, the only possibilities here are (3,3,3), (4,4,4) and (5,5,5).

Case 2. $a = b < c.$

Consider a typical a-sided face $ABCD\ldots$ (see Figure 2.27). Of the other two faces at the vertex B, one must be a-sided, and the other c-sided. Assume without loss of generality that the face alongside BC is c-sided. If we consider the vertex C, it follows that the face alongside CD is a-sided. Thus the neighbors of $ABCD\ldots$ have alternately a sides and c sides. Hence $ABCD\ldots$ must have an even number of neighbors, showing that a is even. Since $3 \leq a \leq 5$, we have $a = 4$. This gives rise to the infinite class $(4,4,n), n \geq 5$.

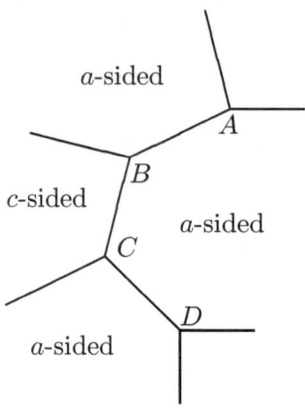

Figure 2.27

Case 3. $a < b = c.$

The characteristic equation in this case is

$$\frac{1}{a} + \frac{2}{b} = \frac{2}{V} + \frac{1}{2}.$$

Recall that $3 \leq a \leq 5$. As in Case 2, b must be even. Suppose $a = 3$. We have $V = \frac{12b}{12-b}$. Since V is a positive integer, the only meaningful values are $b = 4$, 6, 8 and 10. This gives rise to the vertex sequences (3,4,4), (3,6,6), (3,8,8) and (3,10,10). Suppose $a = 4$. We have $V = \frac{8b}{8-b}$. Hence $b = 6$, giving rise to (4,6,6). Finally, suppose $a = 5$. We have $V = \frac{20b}{20-3b}$. Hence $b = 6$, giving rise to (5,6,6).

Case 4. $a < b < c.$

The characteristic equation in this case is

$$\frac{1}{a} + \frac{1}{b} + \frac{1}{c} = \frac{2}{V} + \frac{1}{2}.$$

Here, as in Case 2, all of a, b and c must be even. Since $3 \leq a \leq 5$, we have $a = 4$, and the characteristic equation simplifies to

$$\frac{1}{b} + \frac{1}{c} = \frac{2}{V} + \frac{1}{4}.$$

Suppose $b \geq 8$. Recall $c > b$, so $\frac{2}{V} + \frac{1}{4} = \frac{1}{b} + \frac{1}{c} \leq \frac{1}{4}$. We have a contradiction, so $b = 6$. We have $V = \frac{24c}{12-c}$. Hence $c = 8$ or 10, giving rise to the vertex sequences $(4,6,8)$ and $(4,6,10)$.

Part Two. $n = 4$.

Let the vertex sequence be some permutation of $\{a, b, c, d\}$ such that $a \leq b \leq c \leq d$, where $3 \leq a \leq 5$. The characteristic equation is

$$\frac{1}{a} + \frac{1}{b} + \frac{1}{c} + \frac{1}{d} = \frac{2}{V} + 1.$$

Suppose $a \geq 4$. Then $\frac{2}{V} + 1 = \frac{1}{a} + \frac{1}{b} + \frac{1}{c} + \frac{1}{d} \leq 1$. This is a contradiction, so $a = 3$, and the characteristic equation simplifies to $\frac{1}{b} + \frac{1}{c} + \frac{1}{d} = \frac{2}{V} + \frac{2}{3}$. Suppose $b \geq 5$. Then $\frac{2}{V} + \frac{2}{3} = \frac{1}{b} + \frac{1}{c} + \frac{1}{d} \leq \frac{3}{5}$. This is another contradiction, hence $b = 3$ or 4. We consider the two cases separately.

Case 1. $b = 3$.

Here, the characteristic equation simplifies further to

$$\frac{1}{c} + \frac{1}{d} = \frac{2}{V} + \frac{1}{3}.$$

Suppose $c \geq 6$. Then $\frac{2}{V} + \frac{1}{3} = \frac{1}{c} + \frac{1}{d} \leq \frac{1}{3}$. Again this is a contradiction, so $c = 3, 4$ or 5. If $c = 3$, we have the infinite class $(3, 3, 3, n), n \geq 3$.

We now have $a = b = 3$ and $c = 4$ or 5. We shall first show that the two 3's cannot be consecutive in the vertex sequence. Assuming the contrary, we let the vertex sequence be $(3, 3, c, d)$ and consider a typical triangular face ABC (see Figure 2.28). The vertex A must belong to a triangular face adjacent to ABC. Let this be ABD. Now the vertex C must also belong to a triangular face adjacent to ABC, so either A or B must belong to three triangular faces, which is a contradiction.

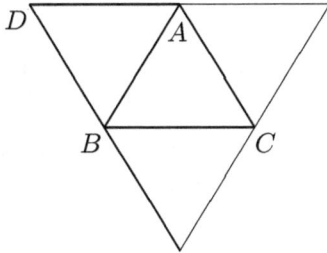

Figure 2.28

We shall now prove that $c = d$, so that the only possible vertex sequences are (3,4,3,4) and (3,5,3,5). Suppose $c < d$. Consider a typical triangular face ABC. The other triangular faces shown in Figure 2.29 are dictated by the form of the vertex sequence, which is $(3, c, 3, d)$. Of the three faces adjacent to ABC, two must contain the same number of edges, and one of A, B or C cannot have $(3, c, 3, d)$ as its sequence, a contradiction.

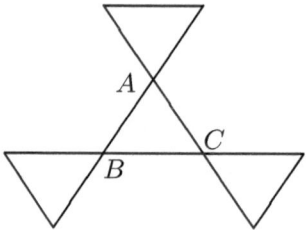

Figure 2.29

Case 2. $b = 4$.

Here, the characteristic equation becomes

$$\frac{1}{c} + \frac{1}{d} = \frac{2}{V} + \frac{5}{12}.$$

Suppose $c \geq 5$. Then $\frac{2}{V} + \frac{5}{12} = \frac{1}{c} + \frac{1}{d} \leq \frac{2}{5}$. This is a contradiction, so $c = 4$. We have $V = \frac{12d}{6-d}$. Hence $d = 4$ or 5. As in Case 1, it can be shown that (3,4,4,5) cannot be a vertex sequence, leaving (3,4,4,4) and (3,4,5,4) as the only possibilities.

Part Three. $n = 5$.

Let the vertex sequence be some permutation of $\{a, b, c, d, e\}$ such that $a \leq b \leq c \leq d \leq e$, where $3 \leq a \leq 5$. The characteristic equation is

$$\frac{1}{a} + \frac{1}{b} + \frac{1}{c} + \frac{1}{d} + \frac{1}{e} = \frac{2}{V} + \frac{3}{2}.$$

Suppose $d \geq 4$. Then

$$\frac{2}{V} + \frac{3}{2} = \frac{1}{a} + \frac{1}{b} + \frac{1}{c} + \frac{1}{d} + \frac{1}{e} \leq \frac{1}{3} + \frac{1}{3} + \frac{1}{3} + \frac{1}{4} + \frac{1}{4} = \frac{3}{2},$$

which shows the supposition to be untenable. Hence $a = b = c = d = 3$. We have $V = \frac{12e}{6-e}$. Hence $e = 3$, 4 or 5, giving rise to the vertex sequences (3,3,3,3,3), (3,3,3,3,4) and (3,3,3,3,5).

This completes the proof of the theorem.

The argument was due to Tom Boag, Charles Boberg and David Hughes [1]. They were junior high school students at the time, when the SMART Circle was not yet in existence. An earlier and different proof of the theorem was given by L. Lines in [3], a book which is now out of print.

The statistics for the Archimedean solids are summarized in the following chart, where F_i denotes the number of faces with i edges. These are obtained by expressing all other variables in terms of V and then substituting into Euler's Formula.

Vertex Sequences	Statistics							
	E	V	F_3	F_4	F_5	F_6	F_8	F_{10}
(3,6,6)	18	12	4	0	0	4	0	0
(3,8,8)	36	24	8	0	0	0	6	0
(3,10,10)	90	60	20	0	0	0	0	12
(4,6,6)	36	24	0	6	0	8	0	0
(5,6,6)	90	60	0	0	12	20	0	0
(3,4,3,4)	24	12	8	6	0	0	0	0
(3,5,3,5)	60	30	20	0	12	0	0	0
(4,6,8)	72	48	0	12	0	8	6	0
(4,6,10)	180	120	0	30	0	20	0	12
(3,4,4,4)	48	24	8	18	0	0	0	0
(3,4,5,4)	120	60	20	30	12	0	0	0
(3,3,3,3,4)	60	24	32	6	0	0	0	0
(3,3,3,3,5)	150	60	80	0	12	0	0	0

Exercises

1. The tetrahedron has 6 edges and the square pyramid has 8 edges. These are the simplest two polyhedra. Intuitively, no polyhedron can have exactly 7 edges. Prove this algebraically using Euler's Formula.

2. (a) With two students and one string, form the skeleton of a tetrahedron.

   (b) With three students and one string, form the skeleton of a standard octahedron.

   (c) With four students and one string, form the skeleton of a cuboid.

3. (a) Draw the Schlegel diagram of one orientation of the snub cube.

   (b) Draw the Schlegel diagram of one orientation of the snub dodecahedron.

Bibliography

[1] Tom Boag, Charles Boberg and David Hughes, On Archimedean solids, *Mathematics Teacher*, **72** (1979) 371–376.

[2] Martin Gardner, *Origami, Eleusis, and the Soma Cube*, Mathematical Association of America, Washington (2008), 1–10.

[3] L. Lines, *Solid Geometry*, Dover Publications Inc., New York, (1965) 159–169.

[4] Hee-Joo Nam, Giavanna Valacco and Ling-Feng Zhu, A Mathematical Performance (I), *Crux Mathematicorum*, **41** (2015) 392–396.

[5] Hee-Joo Nam, Giavanna Valacco and Ling-Feng Zhu, A Mathematical Performance (II), *Crux Mathematicorum*, **41** (2015) 431–434.

[6] Karl Schaffer, A Platonic sextet in strings, *College Mathematics Journal*, **43** (2012) 64–69.

Chapter Three: Polyform Compatibility

Section 1. Tetris Number Theory.

Polyominoes are connected plane figures formed of joining unit squares edge to edge. We have a monomino, a domino, and two trominoes named I and V, shown in Figure 3.1.

Figure 3.1

The five tetrominoes, featured in the popular video game Tetris, are named I, L, S, O and T, respectively. They are shown in Figure 3.2.

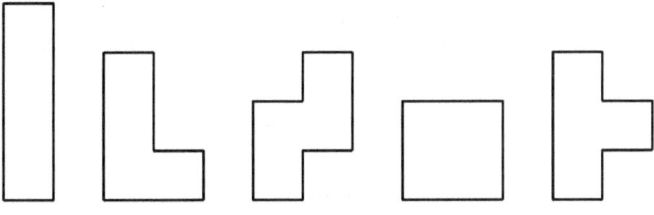

Figure 3.2

A polyomino A is said to **divide** another polyomino B if a copy of B may be assembled from copies of A. We also say that A is a **divisor** of B, B is **divisible** by A, and B is a **multiple** of A. The monomino divides every polyomino.

A polyomino is said to be a **common divisor** of two other polyominoes if it is a divisor of both. It is said to be a **greatest common divisor** if no other common divisor has greater area. Note that we say a greatest common divisor rather than the greatest common divisor since it is not necessarily unique. For instance, the two hexominoes in Figure 3.3 have both the I-tromino and the V-tromino as their greatest common divisors.

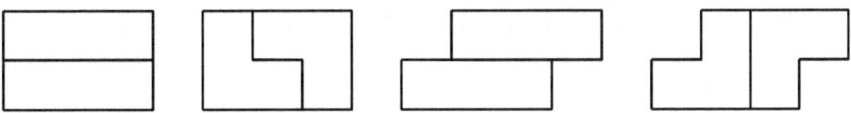

Figure 3.3

© Springer International Publishing AG 2018
A. Liu, *S.M.A.R.T. Circle Projects*, Springer Texts
in Education, DOI 10.1007/978-3-319-56811-9_3

When two polyominoes have at least two greatest common divisors, each divisor is clearly not divisible by any of the others. However, even if a unique greatest common divisor exists, it is still not necessarily divisible by the other common divisors. For instance, the two dodecominoes in Figure 3.4 have the I-tetromino as their unique greatest common divisor, but it is not divisible by the I-tromino which is also a common divisor.

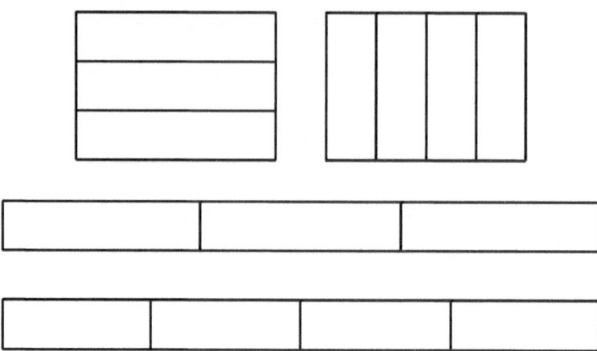

Figure 3.4

Any two polyominoes have a greatest common divisor, since we can always fall back on the monomino. When the greatest common divisor is the monomino, we say that these two polyominoes are **relatively prime** to each other. The monomino is relatively prime to every other polyomino. A **prime** polyomino is one which is divisible only by itself and the monomino, and it is also relatively prime to every other polyomino. Note that the monomino is not considered to be a prime polyomino.

If the area of a polyomino is a prime number, then it must be a prime itself. The converse is not true. The smallest counter-example is the T-tetromino. It has area 4, but is a prime polyomino.

A polyomino is said to be a **common multiple** of two other polyominoes if it is a multiple of both. If two polyominoes have common multiples, they are said to be **compatible**. A **least common multiple** of two compatible polyominoes is a common multiple with minimum area. As shown earlier, the I-tromino and the V-tromino have at least two least common multiples. Clearly, neither multiple divides the other. These two trominoes even have a common multiple, as shown in Figure 3.5, which has an area not divisible by 6, the area of their least common multiple.

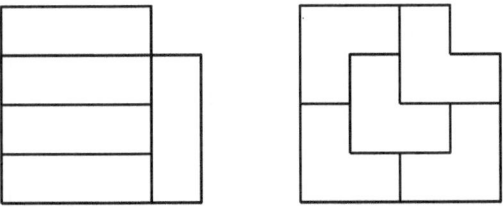

Figure 3.5

However, the area of every common multiple of the I-tromino and the I-tetromino must be a multiple of 12, the area of their least common multiple.

Given two small polyominoes, it is a trivial matter to determine all common divisors of them. It is a different situation with common multiples. To determine whether they are even compatible is often an interesting question. Finding the area of a least common multiple of two compatible polyominoes can also be challenging.

The monomino is trivially compatible with every polyomino. This property is not shared even by the domino, which is incompatible with the icosomino in Figure 3.6. Thus compatibility is not a transitive relation.

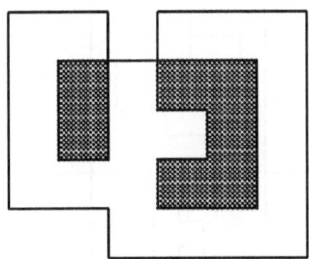

Figure 3.6

We have seen that the two tetrominoes are compatible with each other. They are also compatible with all five tetrominoes. Figure 3.7 shows the constructions of the common multiples with the tetrominoes. The same construction is used for both trominoes except in the case of the T-tetromino.

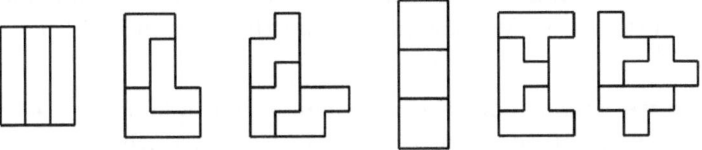

Figure 3.7

Figure 3.8 gives a least common multiple of each pair of tetrominoes. The tetrominoes are featured along the main diagonal. The figure in the i-th row and the j-th column shows how a least common multiple of the i-th and j-th tetrominoes can be constructed from the i-th one.

Figure 3.8

Note that the minimum possible area is attained in all but two cases, between the O-tetromino on the one hand, and the T-tetromino and the S-tetromino on the other. We now justify that these are indeed least common multiples.

Suppose we wish to find a least common multiple of the O-tetromino with either the T-tetromino or the S-tetromino. Clearly, the area of any multiple of a tetromino is a multiple of 4. Since the tetrominoes in question are distinct, the smallest possible area of a common multiple is 8.

Note that two copies of the O-tetromino can abut in essentially two ways as shown in Figure 3.9. Neither figure can be assembled from copies of either the T-tetromino or the S-tetromino. Hence a common multiple has area at least 12.

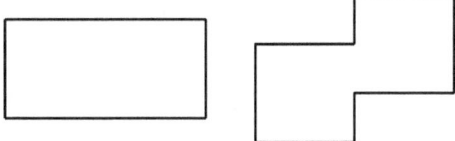

Figure 3.9

If we paint the squares of the infinite grid black and white in the usual checkerboard fashion as shown on the left of Figure 3.10, then three copies of the O-tetromino always cover an even number of white squares while three copies of the T-tetromino always cover an odd number of white squares. Hence they have no common multiples with area 12.

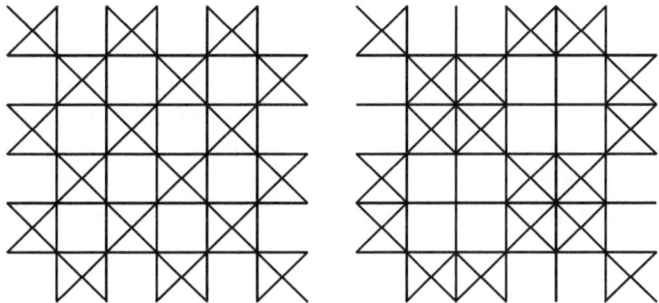

Figure 3.10

If we paint the squares of the infinite grid black and white in the checkered pattern shown on the right of Figure 3.10, then three copies of the O-tetromino always cover an even number of white squares while three copies of the S-tetromino always cover an odd number of white squares. Hence they have no common multiples with area 12 either. It follows that in both cases, the minimum area of a common multiple is indeed 16, as shown in Figure 3.8.

The structure considered in this section is an example of a *Normed Division Domain* considered by Solomon W. Golomb [7]. The presentation is a slight modification of [4]. See also [1] and [2].

Section 2. Tetris Algebra

In *Tetris Algebra*, a different variable such as u, v, w, x or y stands for a different tetromino. A sum of variables means a figure constructed from a specified combination of tetrominoes. Two sums are equal if the figures constructed are identical.

For a sum involving two tetrominoes, there are two cases:

(A) The two tetrominoes are identical.

(B) The two tetromineos are different.

These lead to three situations and five equations.

(A) versus (A):	$2x \ = \ 2y.$	(1)
(A) versus (B):	$2x \ = \ x + y;$	(2)
	$2w \ = \ x + y.$	(3)
(B) versus (B):	$w + x \ = \ w + y;$	(4)
	$x + y \ = \ u + v.$	(5)

In ordinary algebra, each of equations (2) and (4) implies $x = y$. Since two different tetrominoes cannot be identical, it would appear that neither equation can be satisfied. However, the Cancellation Law does not apply in *Tetris Algebra*. As it turns out, one of them has solutions while the other does not. Note that we cannot cancel the common factor 2 from equation (1) either.

When an equation has solutions, our task is to find all possible ones. When it does not, we will have to present a proof of non-existence. A useful tool is to consider various colorings of the infinite grid. We have employed this technique in the last section, with two colorings in Figure 3.10. We shall call them Colorings I and III. The two in Figure 3.11 will be called Colorings II and IV.

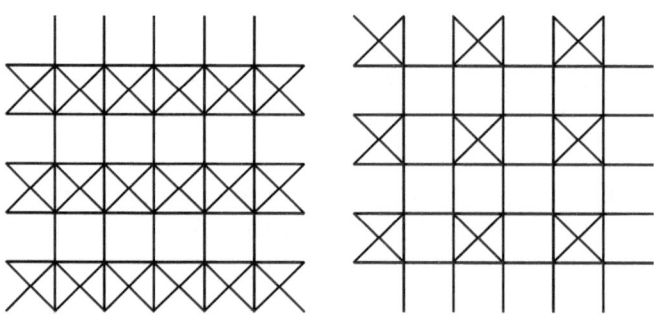

Figure 3.11

In the chart below, we show the possible distributions of black and white squares covered by each tetromino when it is placed on the infinite grid with a specific coloring.

Tetrominoes	Coloring I	Coloring II	Coloring III	Coloring IV
T	**3+1**	3+1 2+2	3+1 2+2	3+1 2+2/4+0
L	2+2	**3+1**	3+1 2+2	3+1 2+2/4+0
N	2+2	2+2	**3+1**	3+1
O	2+2	2+2	2+2/4+0	**3+1**
I	2+2	2+2/4+0	2+2	2+2/4+0

Note the diagonal of entries in boldface. In each case, the number of black squares and the number of white squares covered must be odd. In each entry below this diagonal, the number of black squares and the number of white squares covered must be even. In each entry above this diagonal, both are possible except for the S-tetromino in Coloring IV.

Consider equation (5): $x+y = u+v$. If the T-tetromino is used, the sum with the T-tetromino will cover an odd number of black squares in Coloring I while the other sum will cover an even number of black squares. Hence the two figures cannot be identical. This means that we must use the other four tetrominoes. However, the sum with the L-tetromino will cover an odd number of black squares in Coloring II while the other sum will cover an even number of black squares. Hence the two figures cannot be identical, and equation (5) has no solutions.

Consider now equation (2): $2x = x + y$. As in the above argument, Coloring I eliminates the T-tetromino and Coloring II eliminates the L-tetromino. Now the Coloring III eliminates the S-tetromino and Coloring IV eliminates the O-tetromino. Hence equation (2) also has no solutions.

Below are all possible solutions to equations (1), (3) and (4).

Equation (1): $2x = 2y$.

As we have seen in Section 1, two of the pairs, namely T-O and S-O, are unsolvable, while the other eight cases have solutions.

Equation (3): $2w = x + y$.

Coloring I shows that T can only play the part of w. If w is T, then x and y can be any two of the other four, at least theoretically. In practice, four of the pairs, namely L-O, S-O, S-I and O-I, are unsolvable. If T is not used, Coloring II shows that L can only play the part of w. If w is L, then x and y can be any two of N, O and I. Finally, if L is also not used, Coloring III shows that w must be N and x and y must be O and I. However, Coloring IV shows that there are no solutions. The five solvable cases are shown in Figure 3.12.

Figure 3.12

Equation (4): $w + x = w + y$.

Coloring I shows that T can only play the part of w. If w is T, then x and y can be any two of the other four, at least theoretically. In practice, two of the pairs, namely S-O and O-I, are unsolvable. If T is not used, Coloring II shows that L can only play the part of w. If w is L, then x and y can be any two of N, O and I. Finally, if L is also not used, Coloring III shows that w must be N and x and y must be O and I. However, Coloring IV shows that there are no solutions. The seven solvable cases are shown in Figure 3.13.

Figure 3.13

This section is based on a paper [6] by David Chou and Neo Lin, members of the Chiu Chang Mathematics Circle.

Section 3. Other Compatibility Problems.

The polyominoes are but one special case of polyforms. Another special case consists of the polyiamonds. They are connected plane figures formed of joining unit equilateral triangles edge to edges.

Figure 3.14 shows the moniamond, the diamond, the triamond, three tetriamonds named A, I and V, and four pentiamonds named A, I, J and U.

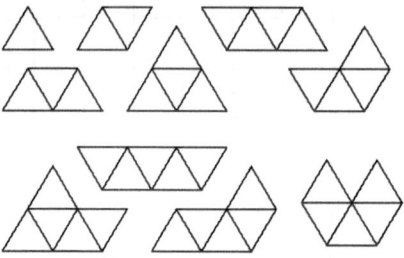

Figure 3.14

We investigate compatibility among the tetriamonds and pentiamonds as well as between the tetriamonds and the pentiamonds.

Figure 3.15 shows that the tetriamonds are compatible with one another.

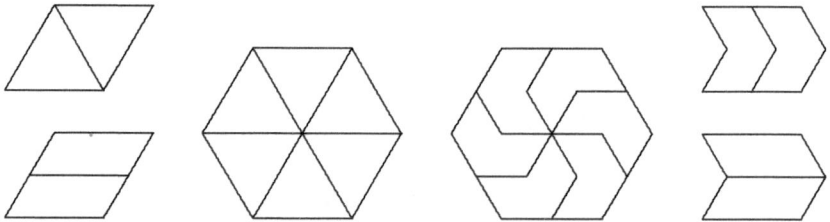

Figure 3.15

The compatibility of the A-Tetriamond with the pentiamonds is the most difficult case. There are no known common multiples of it and the U-Pentiamond. Figure 3.16 shows that it is compatible with the other three pentiamonds.

Figure 3.16

Figure 3.17 shows that the I-Tetriamond is compatible with all four pentiamonds.

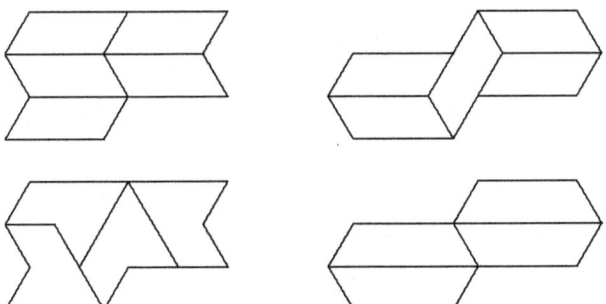

Figure 3.17(a)

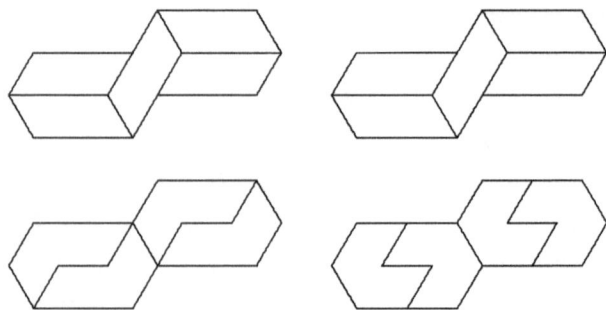

Figure 3.17(b)

Figure 3.18 shows that the V-Tetriamond is compatible with all four pentiamonds.

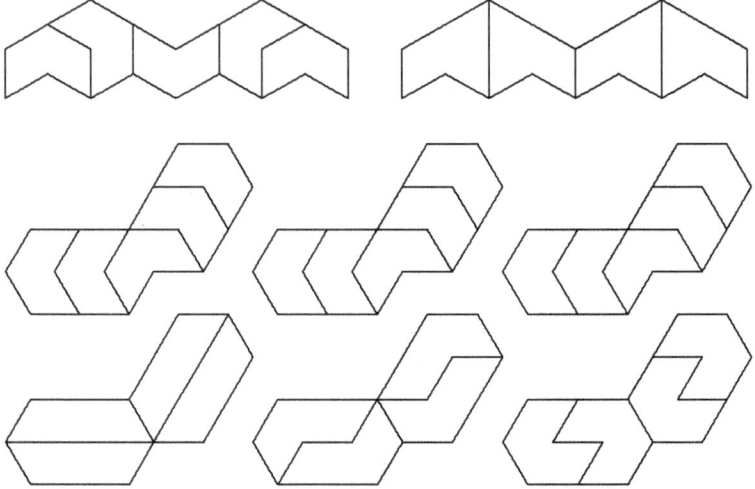

Figure 3.18

Figure 3.19 shows that the pentiamonds are compatible with one another.

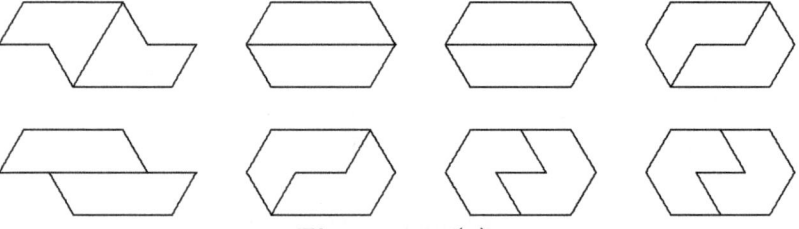

Figure 3.19(a)

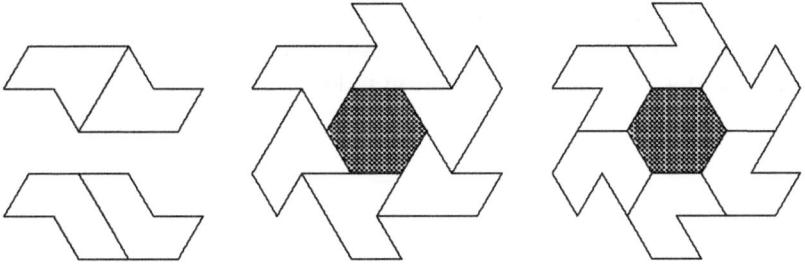

Figure 3.19(b)

The third and last regular polygon which tiles the plane is the hexagon, which gives rise to the polyhexes. Figure 3.20 shows the monohex, the dihex, three trihexes named A, I and V, and seven tetrahexes named I, J, O, P, S, U and Y.

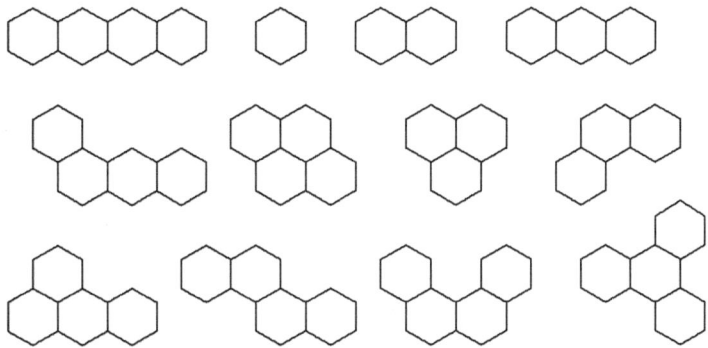

Figure 3.20

Figure 3.21 shows that the trihexes are compatible with one another.

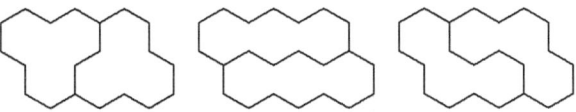

Figure 3.21

Interested readers are invited to extend our investigation to the tetrahexes. We should mention that there are no known common multiples of the I- and Y-Tetrahexes.

This section is based on [3] and [5]. A polyiamond compatibility problem was posed in the Internaitonal Mathematics Competition. In preparation for this contest, Circle members Richard Mah, Ryan Nowakowsky and William Wei carried out a further investigation independent of [3]. Their work was published in [8].

Exercises

1. (a) Find a tetromino such that we can use three copies of it to form a multiple of the V-tromino which contains a completely surrounded 1×1 hole (that is, all eight neighboring squares are part of the figure).

    (b) Are there other tetrominoes with the same property?

2. Find all solutions to solvable equations with each side a sum of two polyominoes, when only the domino and the two trominoes are available.

3. (a) Show that the triamond is compatible with all three tetriamonds.

    (b) Show that the triamond is compatible with all four pentiamonds.

Bibliography

[1] U. Barbans, Andris Cibulis, Gilbert Lee, Andy Liu and Robert Wainwright, Polyomino Number Theory (II), , in *Mathematical Properties of Sequences and other Combinatorial Structures*, edited by J. S. No, H. Y. Song, T. Helleseth and P. V. Kumar, Kluwer, Boston, (2003) 93–100.

[2] U. Barbans, Andris Cibulis, Gilbert Lee, Andy Liu and Robert Wainwright, Polyomino Number Theory (III), in *Tribute to a Mathemagician*, edited by B. Cipra, E. Demaine, M. Demaine and T. Rodgers, A K Peters, Natick, (2005) 131–136.

[3] Andris Cibulis, Andy Liu, M. Lukjanska and George Sicherman, Polyiamond Number Theory, Journal of Recreational Mathematics, **33**-1 (2005) 39–47.

[4] Andris Cibulis, Andy Liu and Robert Wainwright, Polyomino Number Theory (I), Crux Mathematicorum, **28** (2002) 147-150.

[5] Andris Cibulis and George Sicherman, Polyhex Compability, Math Horizons, November (2006) 36–37, 43.

[6] David Chou and Neo Lin, Tetris Algebra, Journal of Recreational Mathematics. 33-3 (2005) 182–192

[7] Solomon Golomb, Normed Division Domains, Amer. Math. Monthly, **88** (1981) 680–686.

[8] Richard Mah, Ryan Nowakowsky and William Wei, Polyiamond Compatibility. Delta-K, **50**-1 (2012) 22–23.

Chapter Four: Mathematical Chess Problems

Section 1. Adventures of an Apprentice Rook

The Pawn Timmy had just been promoted to a Rook, and was being apprenticed to the King's Rook. During his training period, he was instructed to move slowly and cautiously, one square at a time.

"Starting tomorrow, you must visit each of the 64 squares of the chessboard exactly once each day," the King's Rook said. "You may choose to always start at square a1 or always start at square b2. You must take a different path everyday. When all paths have been exhausted, you will have completed your training."

Timmy wished to make his training period as short as possible. Since he had to choose his starting point the next morning, he began counting the number of different paths starting from a1, and the number of different paths starting from b2. It was a difficult task. He was never sure whether he was counting some of them more than once, or whether he had missed some. In any case, the numbers were large enough to put him in despair.

Nevertheless, making the right choice would be a small victory. He drew a diagram of the chessboard and marked square a1 as A and square b2 as B. He also marked square b1 as Y and square a2 as Z. See Figure 4.1. Then he drew a path from A through Y and B to Z.

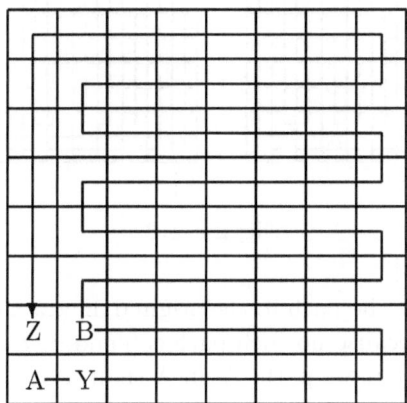

Figure 4.1

This did not help at first. Then Timmy realized that both A and B are black squares. Since the squares along a path must alternate in color, no path can start from A and end at B, or vice versa. This allowed him to see that there must be more paths starting from A than those starting from B.

© Springer International Publishing AG 2018
A. Liu, *S.M.A.R.T. Circle Projects*, Springer Texts
in Education, DOI 10.1007/978-3-319-56811-9_4

Timmy's reasoning is as follows. For each path starting from B, since the path cannot end at A, it must visit A between visits to Y and Z. Suppose the path visits Y first. Then it corresponds to the following one starting from A: move to Y, follow the original path in reverse to B, move to Z, and follow the original path to the end. Figure 4.2 shows an example.

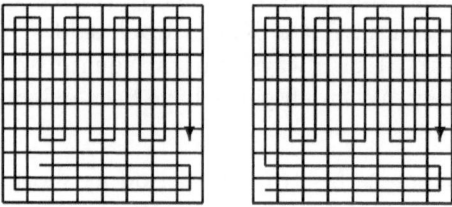

Figure 4.2

If the path visits Z first, then start from A, move to Z, follow the original path in reverse to B, move to Y, and follow the original path to the end. Hence every path from B corresponds to some path from A. Figure 4.3 shows an example.

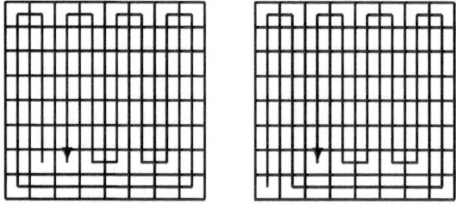

Figure 4.3

On the other hand, the path in the original diagram does not correspond to any path from B because no such path can end at Z unless it moves from A to Z. Thus there are more paths from A than from B.

So the next morning, Timmy chose square b2 as his starting point. The King's Rook said nothing, but Timmy sensed that he had passed his first test.

One day, Timmy found some barriers placed between pairs of adjacent squares. He was able to traverse the entire chessboard without crossing any barriers. However, he realized that had some the barriers been placed differently, his task might not have been possible. He filed a modest protest to the King's Rook.

"The King is trying to turn the chessboard into a labyrinth for the upcoming celebration of the Queen's birthday. Don't worry, I will make sure that it is a *good* labyrinth, that is, one that allows you complete your task."

"Thank you, sir," said Timmy. Then he added, "Surely, there must be more *bad* labyrinths than good ones."

"Why?"

Timy was stuck for an answer. He went away and tried to work things out. He remembered the one-to-one correspondence approach which helped him choose the starting point of his path. He was soon able to justify his somewhat rash statement to the King's Rook.

Timmy's reasoning is as follows. The chessboard has 7 internal vertical grid line, each divided into 8 segments. Hence there are 56 possible vertical barriers. Similarly, there are 56 possible horizontal barriers. Since a path consists of 63 moves, a good labyrinth can have at most 49 barriers. Two labyrinths are said to be complementary if every possible barrier appears in exactly one of them. In any such pair, at least one of them has at least 56 barriers and is bad. Hence there are at least as many bad labyrinths as good ones. Consider the labyrinth with only two barriers isolating a corner square of the chessboard. Then both it and its complement are bad. It follows that there are more bad labyrinths than good ones.

One day, the Queen's Rook came over for a visit. The two Rooks played the following game. The Queen's Rook placed Timmy on any square in column a. Then the two Rooks alternately ordered Timmy to move, the King's Rook issuing the first order. Timmy might not be ordered to move to the left, and not to a square which he had already visited in the game. Eventually, someone had to order him to move to column h, and that player would be the loser.

Timmy was able to figure out a winning strategy for the King's Rook, and told him in a whisper. The King's Rook was very pleased. He followed Timmy's strategy, and won the game.

Timmy's strategy is as follows. If the Queen's Rook places him on square a1, a3, a5 or a7, the King's Rook orders him to move up. If neither player orders Timmy to move to the right, he will be ordered to move to a8 by the King's Rook, after which the Queen's Rook must order Timmy to move to the right. If the Queen's Rook places Timmy initially on square a2, a4, a6 or a8, the King's Rook orders him to move down. In any case, the Queen's Rook must order Timmy to move to column b sooner or later. Then the situation is the same as before, so that the Queen's Rook must lose the game.

The next day, the Queen's Rook brought along his apprentice Tommy, another newly promoted Rook. The King's Rook placed Timmy on one square of the chessboard, and the Queen's Rook placed Tommy on another square. Then they alternately ordered their apprentices to move. What they wanted was that every possible position of the two boys on the chessboard would appear exactly once. They were unable to do so. That evening, Timmy came to the conclusion that what the two Rooks sought was a mission impossible.

His reasoning is as follows. For any position of Timmy, there are 63 possible positions for Tommy. Since they move alternately, each visit of Timmy to that square accounts for 2 of the 63 possible positions for Tommy. Since 63 is odd, this means that Timmy must either end his moves there or begin there. It is clearly not possible for Timmy to do so in every square.

The Queen's Rook came back the next day, and the exercise was repeated, except that Timmy and Tommy did not have to move alternately. They were still unable to make every possible position of the two boys on the chessboard appear exactly once. Again, Timmy saw why this was impossible.

His reasoning is as follows. Define a position as even if both boys are on squares of the same color, and odd otherwise. The positions must necessarily be alternately odd and even. The number of even positions is $2\binom{32}{2} = 992$ while the number of odd positions is $32^2 = 1024$. If all even positions appear, then some odd positions must appear more than once.

Well into his training period, Timmy was taken along by the King's Rook for a Jamboree for Apprentice Rooks. There were so many of them that the chessboard was enlarged to 10×10. Each of them was placed on a different square, and ordered to make one move every minute. They might not change directions until they reached the border of the chessboard, whereupon they would reverse their directions. Miraculously, no two of them ever occupied the same square at any time during the hour-long exercise.

"How many of you were there?" asked the King's Rook.

"I did not count, sir, but there seemed to be lots of us."

"What is the maximum number of apprentice Rooks on the chessboard?"

"Well, sir," Timmy said as he thought things out, "Suppose there were three of us moving along a row or column. By the Pigeonhole Principle, two of us must occupy squares of the same color, and these two must occupy the same square sooner or later."

"What does that mean?" asked the King's Rook.

"That means there are at most two of us moving along a row or column, sir."

"Go on."

"Since there are ten rows and ten columns, sir, there are at most forty of us. Could there have been that many?"

"What do you think?"

"Well, sir, I have to construct a placement for forty of us, and assign an initial direction to each which will preclude simultaneous occupation of the same square by two of us. This is a tall order. For any row or column, although there are only two of us moving along it, there are on the average four of us occupying squares on it at a time."

"Give it a try. You have been making good progress. I am sure you can work it out."

Thus encouraged, Timmy went away to find pencil and paper. Eventually, he came up with Figure 4.4, which showed that there could have been as many as forty apprentice Rooks on the 10×10 chessboard.

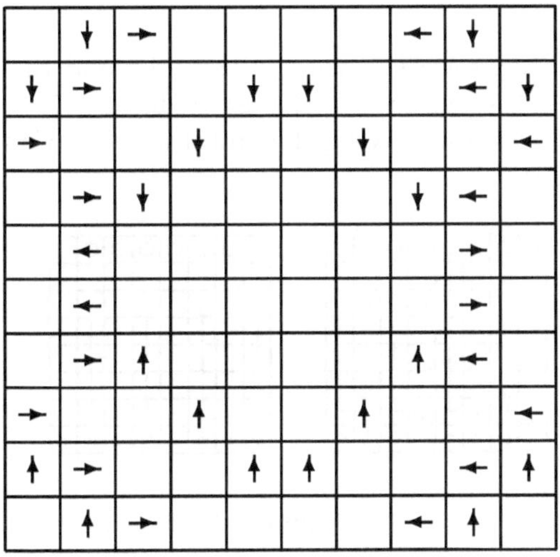

Figure 4.4

The next day, the Jamboree was over, and most of the participants had left. Only sixteen apprentice Rooks were left, and they were given the following exercise on the standard 8×8 chessboard. Eight of them occupied the squares on the top row and the remaining eight occupied the square on the bottom row. They were instructed to move to the opposite edge of the chessboard. At no time could two of them occupy the same square.

Timmy wondered about the minimum number of moves the group must make in order to accomplish the task. Clearly, each of them must make at least 7 vertical moves in order to reach the opposite border. In each column, at least one apprentice Rook must make a horizontal move in order to allow the apprentice Rook coming from the opposite direction to get past. Thus the total number of moves could not be less than $16 \times 7 + 8 = 120$.

It was still necessary to show that 120 moves were sufficient. Timmy constructed a sequence of moves in four stages. He used black circles to represent the apprentice Rooks on the top row and white circles to represent those on the bottom row.

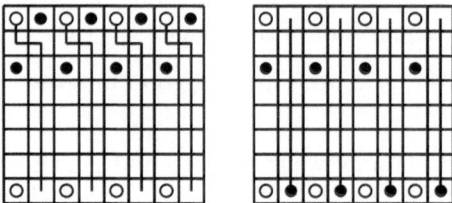

Figure 4.5

The first two stages were shown in Figure 4.5 and the next two stages in Figure 4.6. They required $4 \times (2 + 8) = 40$, $4 \times 7 = 28$, $4 \times 8 = 32$ and $4 \times 5 = 20$ moves respectively. The total number of moves was indeed $40+28+32+20=120$.

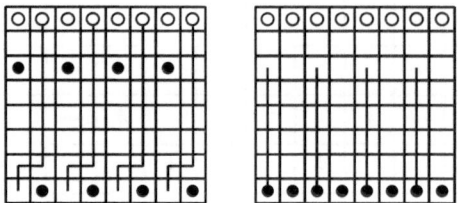

Figure 4.6

"What if the exercise was carried out by eighteen of you on a 9×9 chessboard instead?" the King's Rook, whose approach was not noticed by Timmy while he was concentrating on his deliberation.

"Well, sir," Timmy answered almost immediately. "On a 9×9 chessboard, each of us has to make at least 8 vertical moves. At least one of us from each column must make a horizontal move. So the total number of moves is $18 \times 8 + 9 = 153$."

"This won't work," said the King's Rook.

Timmy thought for a little and said, "I see now, sir. If only 9 horizontal moves are made, there will be five of us in the four even-numbered columns and only four of us in the five odd-numbered columns. So at least one more horizontal move has to be made, bringing the total to 154. I am sure my diagrams may be modified to make this work."

On the final day of his training, Timmy's path ends on a square adjacent to square a1. So he made a 64th move and completed a closed path. He reported this to the King's Rook.

"Could you have made 32 horizontal and 32 vertical moves?"

"I didn't count, sir. It is quite possible."

"It is not," remarked the King's Rook dryly.

Timmy was puzzled. He was the one moving, and he could not remember whether he had made 32 horizontal and 32 vertical moves. How could the King's Rook tell that this was impossible? Timmy remained puzzled for the whole day, until he finally realized that the King's Rook was right, as usual.

Timmy's reasoning is as follows. Each time he moves from one square to the next one, connect the centers of these squares by a segment of length 1. Thus we have a closed path L_0 of length 64. Note that this path encloses a region which does not contain a complete square of the chessboard, since the center of each square is on the path. Figure 4.7 shows a typical closed path.

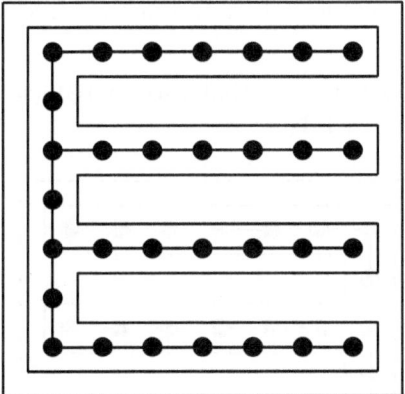

Figure 4.7

Consider all the vertices of the squares of the chessboard that are inside this region as vertices of a graph. If two of them are adjacent vertices of the same square, connect them by the corresponding edge of that square. If this graph has a cycle, it would enclose at least one complete square of the chessboard, and this square would be inside the region enclosed by L_0.

Since this is not the case, the graph is a tree and has a vertex of degree 1. In Figure 4.7, the vertex at the bottom right corner is such a vertex. We remove it and shorten the closed path to L_1 of length 62. Now two squares are not visited, and we glue them together to form a vertical domino, as shown in Figure 4.8.

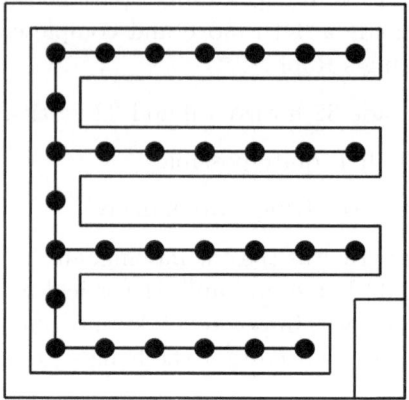

Figure 4.8

Note that L_1 has two horizontal segments less than L_0 but the same number of vertical segments. If the domino is horizontal instead, L_1 will have two vertical segments less than L_0 but the same number of horizontal segments.

Now L_1 still encloses tree, so that further reduction can be made. In Figure 4.9, we reach L_9 with length 48.

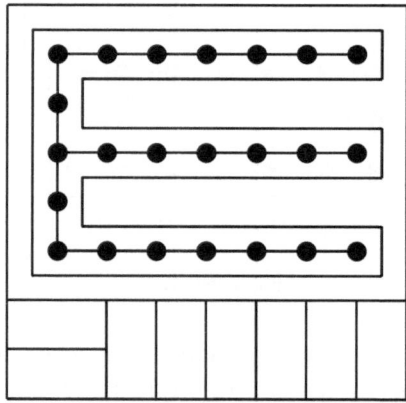

Figure 4.9

Eventually, we reach L_{30} which consists of four segments forming a unit square, with the rest of the chessboard partitioned into 30 dominoes, as shown in Figure 4.10.

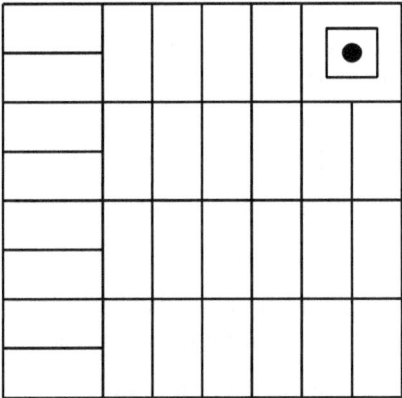

Figure 4.10

Suppose L_0 has 32 horizontal segments and 32 vertical ones. Then 15 of these 30 dominoes are vertical and 15 of them are horizontal. We divide the 2×2 square covered by L_{30} into 2 horizontal dominoes. They are either both horizontal or both vertical. By symmetry, we may assume that they are both horizontal. Then the whole chessboard has been partitioned into 17 horizontal and 15 vertical dominoes. It means that some horizontal grid line of the chessboard must cut across an odd number of vertical dominoes. This is a contradiction since the part above and the part below this grid line outside of the cut dominoes both contain an odd number of squares, and cannot be covered by un-cut dominoes. It follows that the number of horizontal segments cannot be equal to the number of vertical segments.

Although it was well past midnight, Timmy was so excited that he roused the King's Rook out of bed. He listened to Timmy patiently, and then nodded. "Excellent, Timmy!" he said. "Congratulations! You are now a full-fledged Rook!"

Section 2. Martin Gardner's Royal Problem

The Red Queen was furious, as usual. Her current ire was brought on by the absence of the Red King from his Palace. On her rare visits, she expected to see whom she had come to see.

"Bring the old fool back here, or else!" roared the Red Queen, who was related to the Queen of Hearts.

"Or else what?" asked Alice, but only after Her Majesty had swept radiantly out of earshot back to her side of the Palace.

"Off with your head!" Tweedledum said.

"What else?" added Tweedledee rhetorically.

"Oh, dear," said Alice, "this puts a new meaning to ten percent off the top. What shall I do? I don't even know where the Red King is."

The twins brought out a map of the land. It was the familiar 8 × 8 chessboard in Figure 4.11.

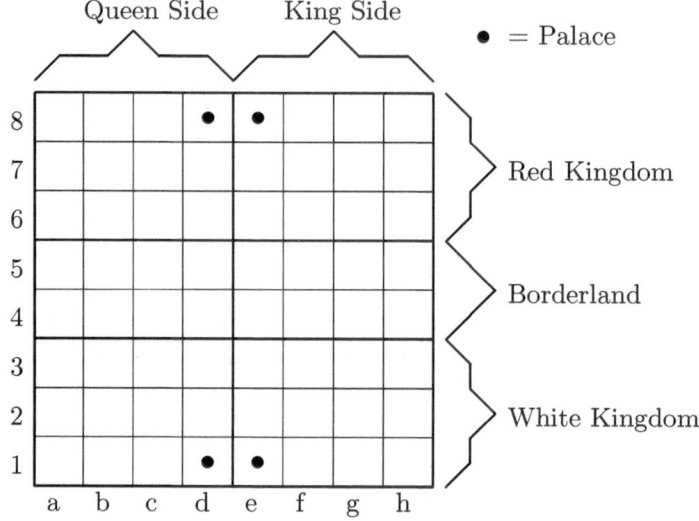

Figure 4.11

"I bet I know where His Majesty is," said Tweedledum.

"On h4!" exclaimed Tweedledee.

"How do you know that?" Alice asked.

"Well," said Tweedledum, "the Red King plays it safe. He never ventures into the White Kingdom."

"He also refuses to cross over to the Queen Side," added Tweedledee.

"So he is confined to twenty squares in Figure 4.12. That is helpful, but I still don't see how you can be so sure that he is on h4."

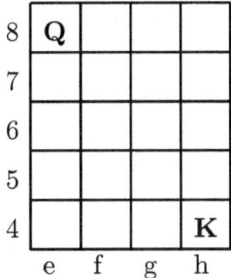

Figure 4.12

"His Majesty likes to be as far away from the Red Queen as he possibly can," Tweedledum said.

"Actually, as far from the Red Queen's Palace as possible," corrected Tweedledee. "He has no control over the whereabouts of Her Majesty."

"There is another problem," said Alice. "If the Red King does not want to come back to e8, how can I persuade him against his wish?"

The twins thought for a while, and fought for a while just to pass the time. Then they both came up with a brilliant idea. Not surprisingly, it was the same idea.

"Are you in mortal fear of the Red Queen?" Tweedledum asked Alice.

"Of course. Who isn't?"

"Of all people, who fears her the most?" asked Tweedledee.

"Hard to say," Alice replied. Then it occurred to her. "The Red King, of course."

"Right!" said Tweedledum. "He could not risk getting caught in a mating situation with the White Queen."

"So if you disguise yourself as that good lady, you can drive His Majesty back here," declared Tweedledee triumphantly.

"It is worth a try," said Alice, somewhat encouraged. "I should not waste any time by venturing outside of those twenty squares either."

"Make sure you do not corner His Majesty on h8 with no legal moves," Tweedledum advised Alice.

"Also, do not drive him into the White Kingdom," said Tweedledee. "His Majesty may find out that it is not as dangerous as he makes it out to be."

"Well, I'd better hurry and bring His Majesty back as soon as I can. The Red Queen's patience is shorter than her temper!"

We summarize Alice's strategy in the following chart.

Moves	(1)	(2)	(3)	(4)	(5)	(6)	(7)
White	e5	f6	f4	e5	f5	e6	h6
Red	g4	h5	g6	h6	g7	h8	g8
Notes				(a)		(b)	
Moves	(8)	(9)	(10)	(11)	(12)	(13)	(14)
White	g6	g5	f6	h6	g5	g6	f5
Red	h8	h7	g8	f7	f8	e7	e8
Notes	(c)				(d)		

Notes:
(a) If (4) ... f7, continue from (12). If (4) ... h7, continue from (10).
(b) If (6) ... f8, continue from (13). If (6) ... h7, continue from (10).
(c) If (8) ... f8, then (9) h7 e8.
(d) If (12) ... e6, then (13) f4 e7, and continue from (14).

Alice drove the Red King back to his Palace just in time.

"Come along," roared the Red Queen. "We have to attend a summit conference with the White Queen and her consort."

"What is the matter this time, dear?" asked the Red King timidly.

"We have been discussing the partition of the Borderland. There is too much goings-on here, especially on h4, or so I hear."

"I can't imagine what," murmured the Red King.

"Anyway, the White Queen and I have agreed to establish our borders between ranks 4 and 5. We just meet to formalize the deal."

"If you say so, dear."

As soon as the new treaty was signed, the Red King headed for h5, the furthest haven within his domain. Alice was dispatched after him once again. As it turned out, her efforts were futile.

We generalize the problem to an $m \times n$ chessboard, $m \geq n \geq 3$. A White Queen is on square $(1,1)$, while a Red King is on square (m,n). The Queen moves first unless $m = n$, in which case the King must move out of check. Thereafter, moves alternate. The Queen wins if and only if the King is forced to her initial square $(1,1)$ in a finite number of moves. With perfect play, the King wins if and only if $m = n$. In our analysis, all positions are considered at the moment when it is the King's turn to move.

We first prove that the King wins if $m = n \geq 3$. We consider the game from his point of view and define a "forbidden zone" into which he must not move. This zone consists of all squares (i, j) where $i + j \leq n - 1$. It always contains the forbidden square $(1,1)$. For $n = 3$, it consists only of this square. The case $n = 4$ is shown in Figure 4.13, with the squares in the forbidden zone marked by black dots.

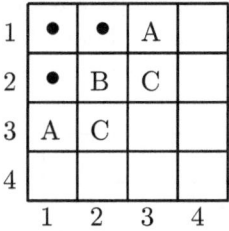

Figure 4.13

We'll prove that not only can the King avoid going to $(1,1)$, he can't even be forced into the forbidden zone by the Queen. For this to happen, the King must be on one of the following types of squares:

A. $(n - 1, 1)$ or $(1, n - 1)$;
B. (i, j), where $i + j = n$, with $i > 1$ and $j > 1$;
C. (i, j), where $i + j = n + 1$, with $i > 1$ and $j > 1$.

For $n = 4$, the relevant squares are marked accordingly in Figure 4.13.

Consider Case A. Suppose the King is on $(n - 1, 1)$ as shown in Figure 4.14. He can move to $(n - 2, 2), (n - 1, 2), (n, 1)$ or $(n, 2)$, none of which is in the forbidden zone. These moves are marked with ×'s in Figure 4.14.

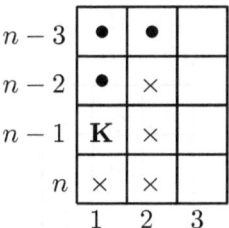

Figure 4.14

The only squares from which the Queen can control all four squares are $(n - 1, 1), (n - 1, 2)$ and $(n, 2)$. The first is already occupied by the King. If the Queen is on either of the other two squares, she will be captured. So the King can't be forced into the forbidden zone when he is on $(n-1, 1)$ or, by symmetry, on $(1, n - 1)$.

Cases B and C can be handled similarly, the King having even more options. This completes the proof that the King wins if $m = n \geq 3$.

While this proof is very simple, one may well ask how we came to think of the forbidden zone in the first place. Our initial approach is by mathematical induction on n. It's not difficult to see that the King wins if $n = 3$.

For $n = 4$, we mark off two overlapping 3×3 boards on the 4×4 board, as shown in Figure 4.15. Each smaller board has its own forbidden square, and the two join up with the actual forbidden square to form the forbidden zone in Figure 4.13.

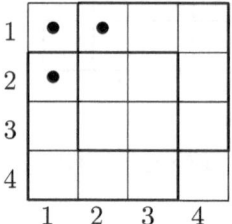

Figure 4.15

We now consider two cases. If the King and Queen are on the same 3×3 board, we already know that the King has a safe square within the same board. If the King and Queen aren't on the same 3×3 board, the Queen is too far away to restrict the King's movement effectively. It's easy to see how the general inductive step goes. We omit the details because our simplified proof makes mathematical induction unnecessary here.

To complete the justification of our claim, we give a winning algorithm for the Queen if $m > n \geq 3$. We consider the game now from her point of view. She will win if she can achieve the position in Figure 4.16, with the King on (i, j), provided that $i + j \leq n$.

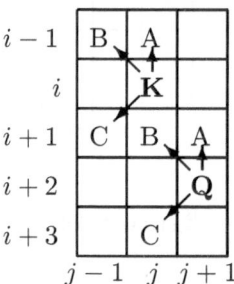

Figure 4.16

From this position, all possible moves by the King are indicated by arrows. The Queen's responses are shown by arrows with matching labels.

Note that after each move, the position is again that in Figure 4.16. The King's column number never increases, and it can't remain constant forever. Thus, the King will be driven to column 1 eventually. It's now a simple matter for the Queen to march him up column 1 to $(1,1)$.

We've already proved that the Queen can't win on a square board. This is a good place to pause and see why the squareness of the board makes such a big difference. It's certainly possible for the Queen to get the King into the position in Figure 4.16, with the King on (i, j), where $i + j \leq n$. It's also possible for the King to keep $i + j = n$. By choosing option C every time, he will reach column 1 on $(n - 1, 1)$. The Queen must now move to $(n + 1, 2)$, but this is possible if and only if $m > n$.

We now prove that the Queen can win, with or without getting the King into the position in Figure 4.16. Her initial objective is to achieve any of the three positions shown in Figure 4.17.

If $n > 5$, this is easily accomplished by the Queen giving check on $(m, 1)$. The King can move to either $(m - 1, n - 1)$ or $(m - 1, n)$. The Queen then goes to $(m, n - 3)$ or $(m, n - 2)$ accordingly. For $n = 4$, the Queen first moves to $(m - 1, 1)$. For $n = 3$, the Queen first gives check on $(m - 2, 1)$. In each of these two special cases, at most two more moves will lead to a desired position. The reader can work out the details.

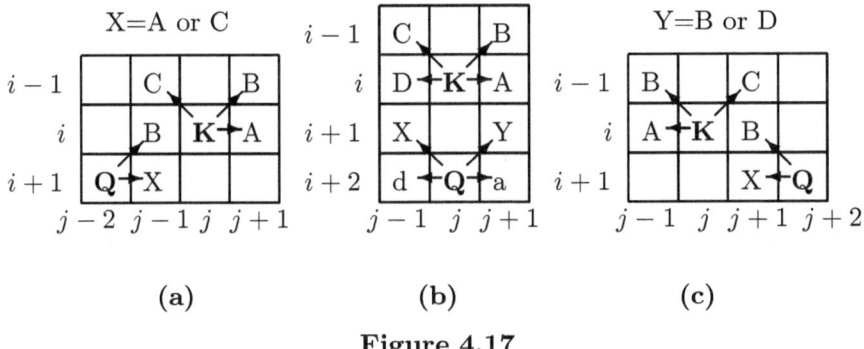

Figure 4.17

From the positions in Figure 4.17, all possible moves by the King are indicated by arrows. The Queen's responses are indicated by arrows with matching labels. If the King happens to be in column 1 or n, option A or D for the Queen in Figure 4.17(b) is impossible. She should make the alternate response with a matching label in the lower case.

Note that after each move, the position is again one of the three in Figure 4.17. The King's row number never increases, and it can't remain constant forever. So the King will be driven to row 1 eventually.

When the King reaches row 1 at $(1, j)$, the Queen abandons the strategy indicated in Figure 4.17. Instead she gives check at $(3, j)$, which she can always do. We now consider three cases.

Case 1: $j < n$ and the King moves to $(1, j - 1)$.
The Queen moves to $(2, j + 1)$ and marches the King along row 1 to $(1,1)$.

Case 2: $j < n$ and the King moves to $(1, j + 1)$.
The Queen continues to check along row 3. If the King moves back toward $(1,1)$ before reaching $(1, n)$, the Queen can convert the situation to that in Case 1. If the King goes to $(1, n)$, the Queen makes an unexpected move, from $(3, n - 1)$ to $(4, n - 1)$. The King's moves are now forced: King to $(2, n)$, Queen to $(3, n - 2)$, King to $(1, n - 1)$, and Queen to $(3, n)$. She has now achieved the winning position in Figure 4.16, since $1 + (n - 1) = n$.

Case 3: $i = n$. The King must move to $(1, n - 1)$. The Queen gives check at $(3, n - 1)$. The King's response will lead to either Case 1 or Case 2.

This completes the proof that the Queen wins if $m > n \geq 3$.

Section 3. Knight Tours

A *knight tour* on a chessboard is a sequence of moves of a knight which takes it over very square of the chessboard. If the tour ends on a square which is a knight's move away from the starting square, the tour is said to be *re-entrant*.

The problem of knight tours is very old. For popular accounts, see [1] and [6]. However, these references describe methods of construction which are not supported by arguments that they will never fail. It is not until recently that constructions with proofs are supplied. See [2] and [8], both of which deal with rectangular boards, and [5], which deals with more general boards.

The aim of our investigation is to find re-entrant knight tours on an $n \times n$ chessboard for even n. Our overall plan of attack is to classify the $n \times n$ boards according to the congruence class of n modulo 6. Thus there are three cases, $n \equiv 0 \pmod 6$, $n \equiv 2 \pmod 6$ and $n \equiv 4 \pmod 6$.

Figure 4.18 shows a re-entrant knight tour on the 6×6 board called T6A. Henceforth, all re-entrant tours will be identified by A.

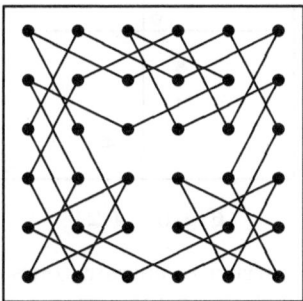

Figure 4.18

Figure 4.19 shows a tour T6B which is not re-entrant. It will be used as our basic building block. Henceforth, copies of T6B will not be labeled.

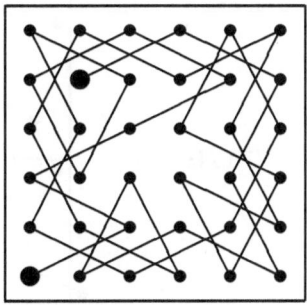

Figure 4.19

Figure 4.20 shows that if we put together 4 copies of T6B, we have T12A! The same idea works for all $12k \times 12k$ boards.

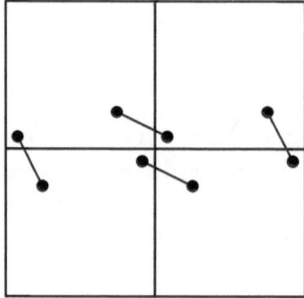

Figure 4.20

In investigating the 18×18 board, we first construct T12B by putting together 3 copies of T6B and 1 copy of T6A, as shown in Figure 4.21.

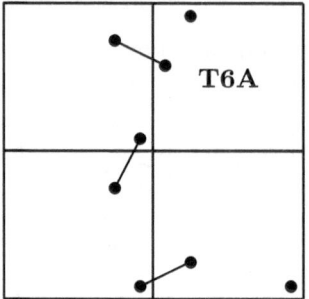

Figure 4.21

This is then used to construct re-entrant tours on all $(12k+6) \times (12k+6)$ boards. T18A is shown in Figure 4.22.

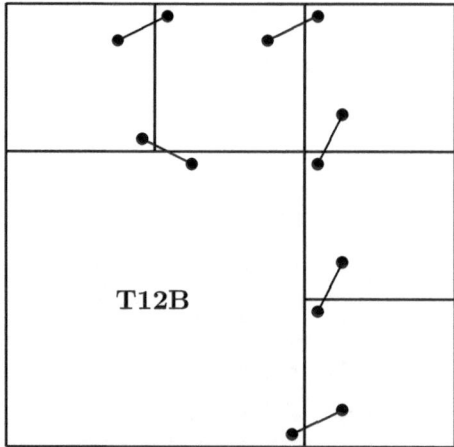

Figure 4.22

We turn to the next case, where $n \equiv 2 \pmod 6$. Clearly, there are no re-entrant tours on the 2×2 board. Many on the 8×8 board are known. After all, this is the standard chessboard, and the problem has been around for a long time. We call the one in Figure 4.23 T8A.

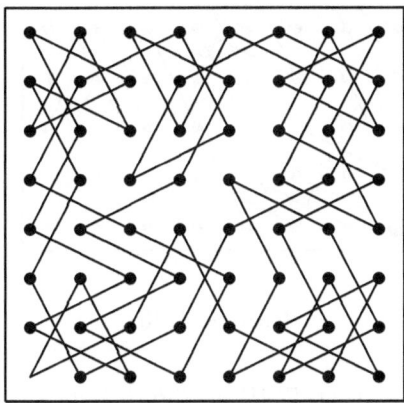

Figure 4.23

Figure 2.24 shows a tour T8B which is not re-entrant.

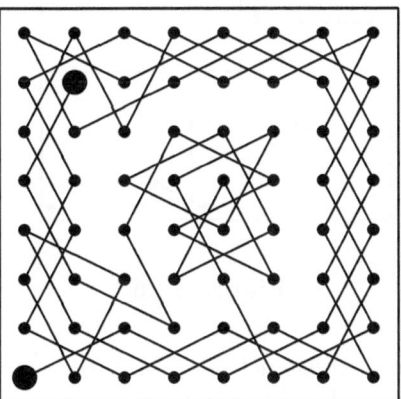

Figure 4.24

Figure 2.25 shows a board which not rectangular but elbow-shaped, with a tour T8C on it. Henceforth, all tours on elbow-shaped boards will be identified by C.

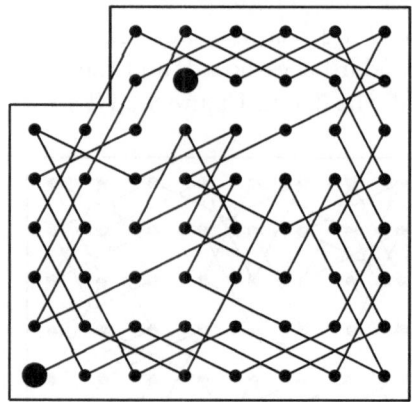

Figure 4.25

Using these, we construct T14B in Figure 4.26.

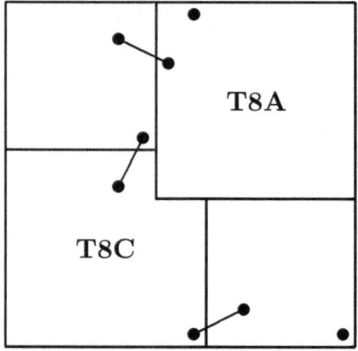

Figure 4.26

This is in turn used to construct re-entrant tours on all $(12k+8) \times (12k+8)$ boards. T20A is shown in Figure 4.27.

Using T8B and T8C, we can get T14A as shown in Figure 4.28, and T26A as shown in Figure 4.29. The method works for all $(12k + 2) \times (12k + 2)$ boards.

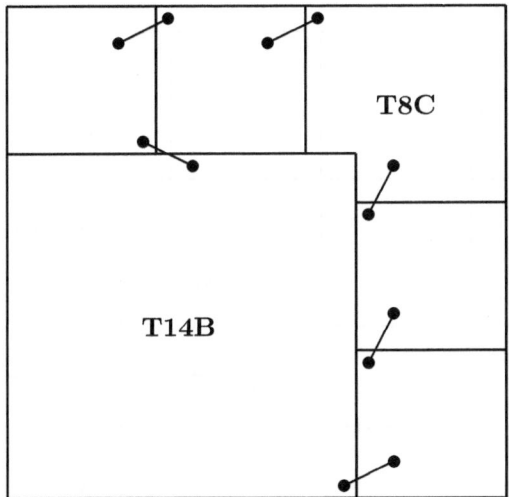

Figure 4.27

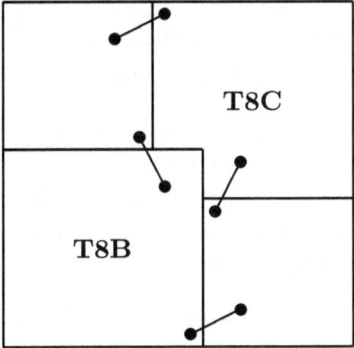

Figure 4.28

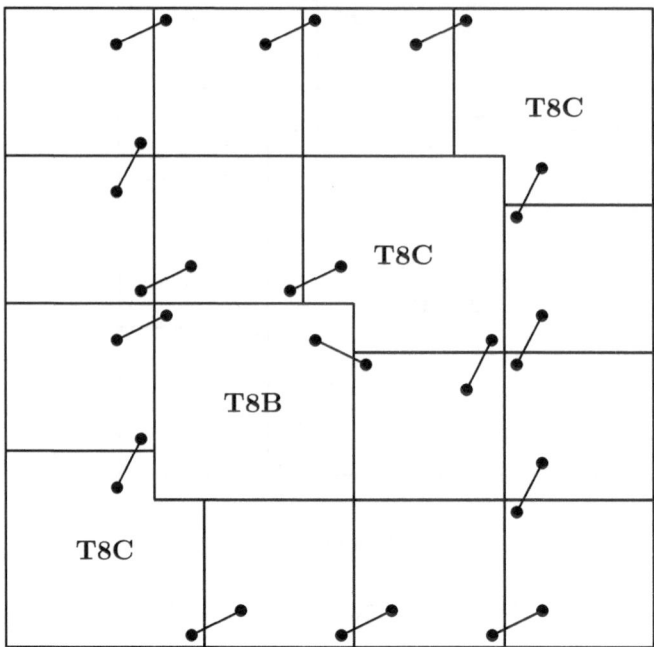

Figure 4.29

We now come to the last case, where $n \equiv 4 \pmod 6$. It is easy to show that there are no re-entrant knight tours on the 4×4 board. This is the only exception. The general construction is analogous to that in the case $n \equiv 2 \pmod 6$, but using instead the pieces T10A in Figure 4.30, T10B in Figure 4.31 and T10C in Figure 4.32.

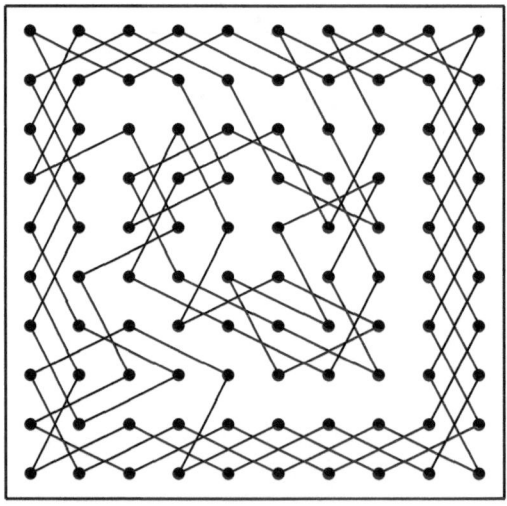

Figure 4.30

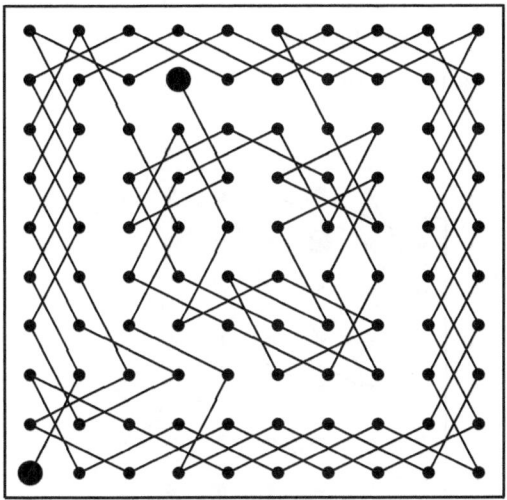

Figure 4.31

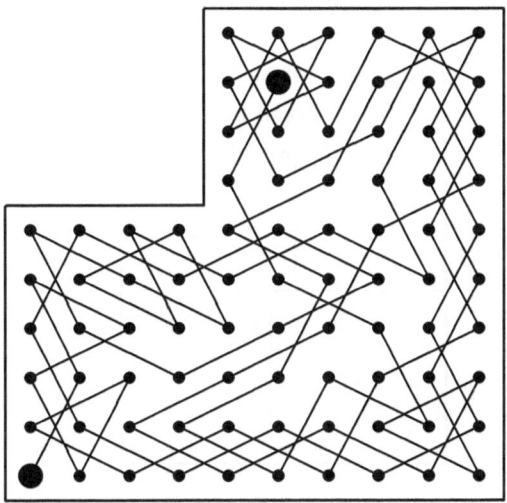

Figure 4.32

In summary, re-entrant knight tours exist on $n \times n$ boards for all even $n \geq 6$.

When n is odd, a re-entrant tour is not possible on the $n \times n$ board. If we paint the squares black and white in the usual chessboard fashion, the numbers of white squares and of black squares differ by 1. Now the Knight always moves between squares of opposite colors. Hence the tour starts and ends on squares of the same color, and they cannot be a Knight's move apart. In a sense, this makes our task easier since we just have to find any knight tour.

Continuing our overall plan of attack, we also have three cases here, $n \equiv 1 \pmod 6$, $n \equiv 3 \pmod 6$ and $n \equiv 5 \pmod 6$.

T1B a trivial knight tour on a 1×1 board. Figure 4.33 shows an elbow-shaped T7C.

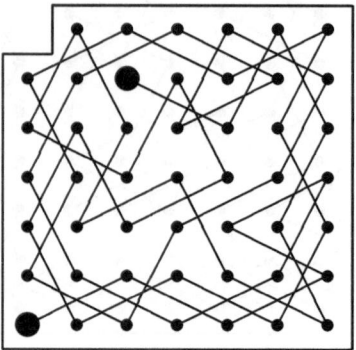

Figure 4.33

T1B and T7C combine easily into T7B. T13B is constructed using T7B, T7C and two copies of T6B, as shown in Figure 4.34.

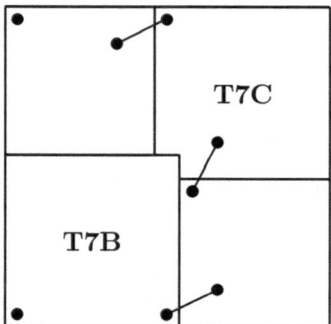

Figure 4.34

The same construction can be used to get a tour on all $n \times n$ boards, $n \equiv 1 \pmod 6$. T19B is shown in Figure 4.35.

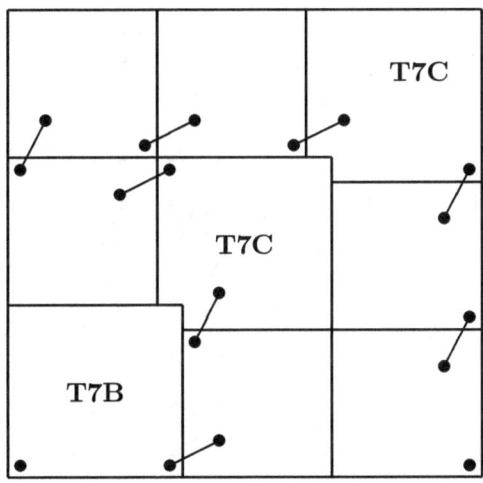

Figure 4.35

It is now clear that the cases $n \equiv 3 \pmod 6$ and $n \equiv 5 \pmod 6$ can be handled in the same way. There are no knight tours on the 3×3 board since the central square is isolated. Apart from this exception, knight tours exists on all other boards of odd sizes. All that remains to be done is to construct T9B, T9C, T5B and T11C. These are shown in Figures 4.36, 4.37, 4.38 and 4.39, respectively.

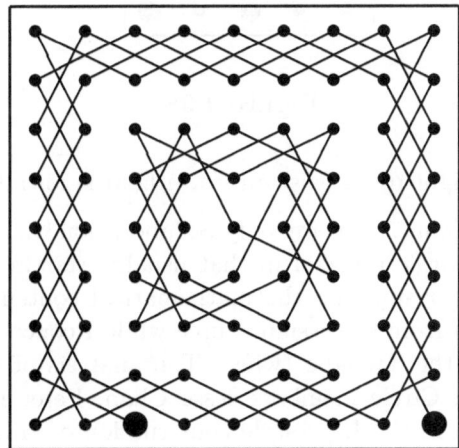

Figure 4.36

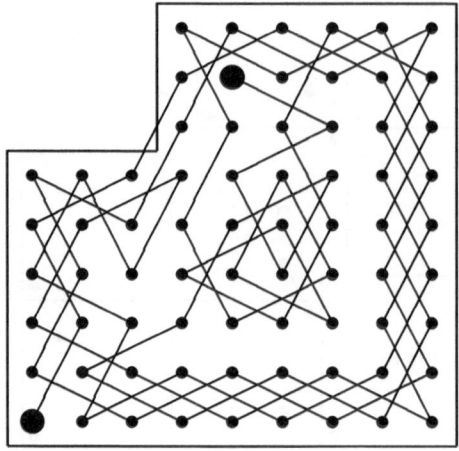

Figure 4.37

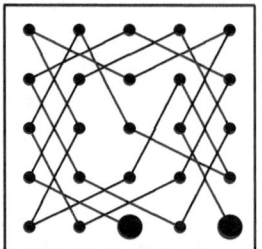

Figure 4.38

In summary, knight tours exist on $n \times n$ boards for all odd $n \neq 3$.

The Royal Problem [7] was my only collaboration with Martin Gardner. There was a sort of role reversal in that usually Martin did the narrative while his collaborator supplied the mathematical contents. Here, Martin dug up the problem from a Russian source while I concocted the exchange between Alice and the Tweedle twins. The analysis of the square board was the work [4] of Circle members Jesse Chan, Peter Laffin and Da Li. The novel approach in the Lego-style construction of knight tours was the work [3] of Circle members Hubert Chan, Steven Laffin and Daniel van Vliet. Hubert and Steven are the younger brothers of Jesse and Peter, respectively.

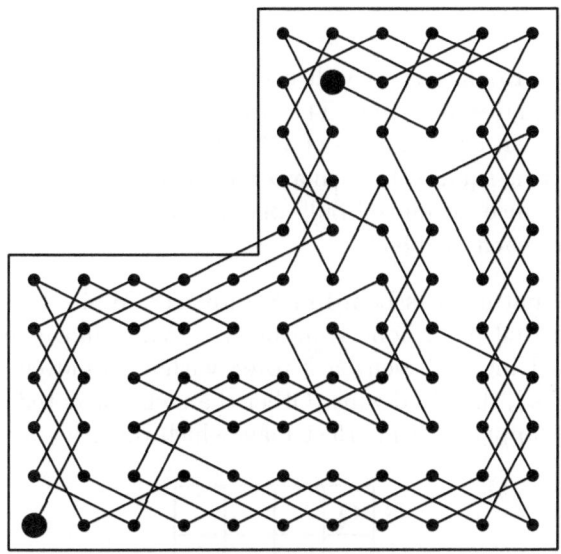

Figure 4.39

Exercises

1. The King's Rook was so well established that he always moved at least two squares at a time. The King ordered him and his apprentice Timmy to visit every square of a 4 × 4 board exactly once and return to the starting square. In such a confined space, Timmy had a decided advantage since he moved only one square at a time. Was the task possible for each of them?

2. On the miniature chessboard in Figure 4.18, White has a lone Queen on e8 and Red has a lone King on h6. White moves first, and wins if the Red King is driven back to e8 within 10 moves. If this is not accomplished, then Red wins. Other than what is noted above, normal chess rules apply. With perfect play, which royalty wins?

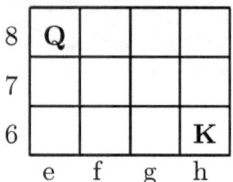

Figure 4.18

3. Prove that there are no re-entrant knight tours on the 4 × n board.

Bibliography

[1] W. W. Rouse Ball and H. S. M. Coxeter, *Mathematical Recreations and Essays*, Dover Publications Inc., Mineola (1978) 175–186.

[2] R. Cannon and S. Dolan, The knight's tour, Mathematical Gazette **70** (1986) 91–100.

[3] Hubert Chan, Steven Laffin and Daniel van Vliet, Knight Tours, Mathematics and Informatics Quarterly **2** (1992) 135–150.

[4] Jesse Chan, Peter Laffin and Da Li, Martin Gardner's Royal Problem, Quantum **4** (1993) 45-46.

[5] R. B. Eggleton and A. Eid, Knight's circuits and tours, Ars Combinatoria. **17A** (1984) 145–167.

[6] Martin Gardner, *Mathematical Magic Show*, Mathematical Association of America, Washington (1990) 188–202.

[7] Martin Gardner and A. Liu, A Royal Problem, Quantum **3** (1993) 30–31.

[8] A. J. Schwenk, Which rectangular chessboards have knight's tours? Mathematics Magazine **64** (1991) 325–332.

Chapter Five: Mathematical Induction

Section 1. The Towers of Hanoi

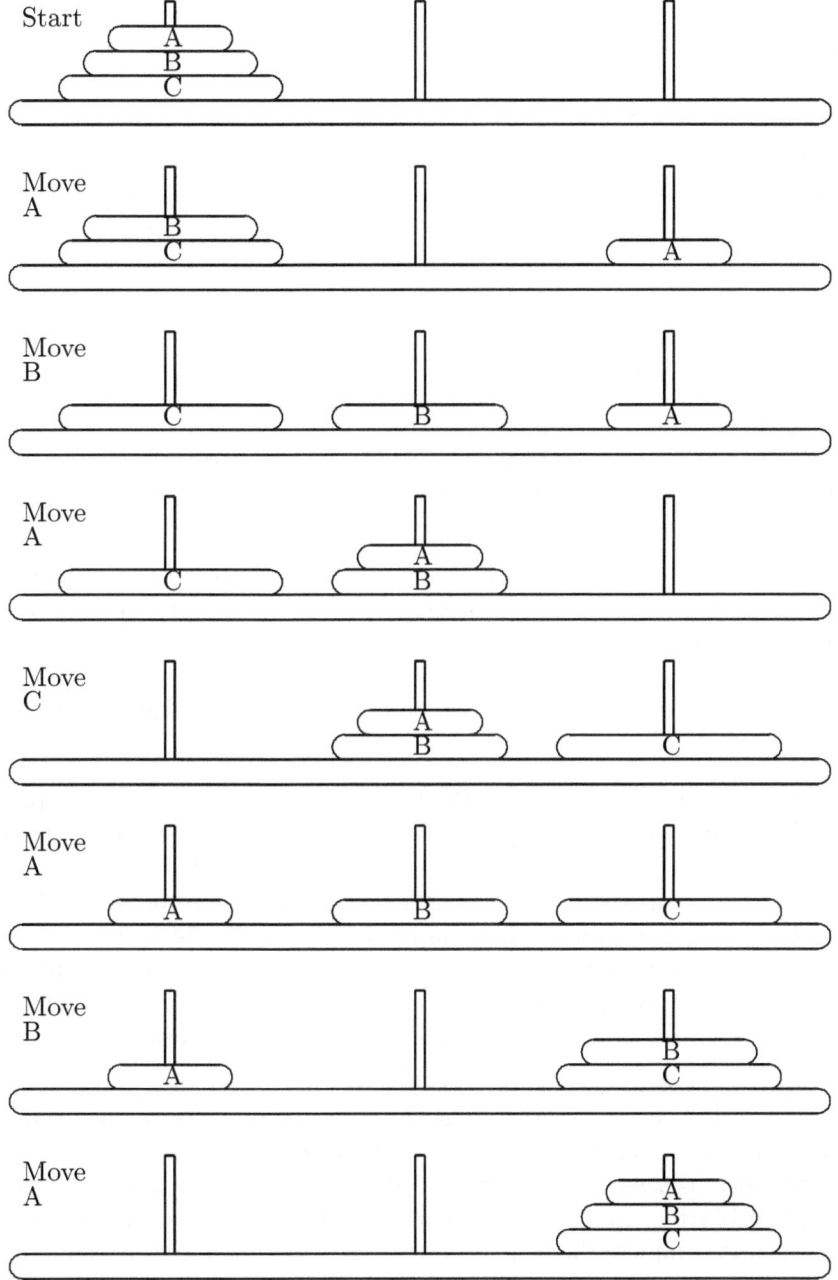

Figure 5.1

© Springer International Publishing AG 2018
A. Liu, *S.M.A.R.T. Circle Projects*, Springer Texts
in Education, DOI 10.1007/978-3-319-56811-9_5

In a famous puzzle known as the Tower of Hanoi, there are three pegs in the playing board. There are several disks of different sizes, all stacked on the first peg, in ascending order of size from the top. The objective is to transfer this tower to the third peg. The rule is that we may only move a disk on top of a peg to the top of another peg, and a disk may not be placed on top of a smaller disk.

If the number of disk is 1, the task can be accomplished in 1 move. If the number of disks is 2, the task can be accomplished in 3 moves. Figure 5.1 shows that if the number of disks is 3, the task can be accomplished in 7 moves: ABACABA. Based on these three simple cases, we conjecture that the minimum number of moves required to transfer a tower with n disks is $2^n - 1$.

To prove this, we turn to a powerful general method.

Principle of Mathematical Induction
Let $P(n)$ be a sequence of statements such that $P(k)$ is true for some positive integer k, and $P(n + 1)$ is true whenever $P(n)$ is true. Then $P(n)$ is true for *all* integers $n \geq k$.

Usually, we have $k = 1$, and wish to conclude that $P(n)$ is true for all positive integers n. A proof by mathematical induction consists of the following two steps.

(1) **Basis.** Prove that $P(k)$ is true for some positive integer k. Usually, we have $k = 1$.

(2) **Induction.** Assuming that $P(n)$ is true for *some* integer $n \geq k$, prove that $P(n + 1)$ is also true. $P(n)$ is called the **induction hypothesis**.

Let us carry out the proof by mathematical induction for our problem. Here, $P(n)$ is the statement that the minimum number of moves required to transfer a tower with n disks is $2^n - 1$. We have already established $P(1)$, $P(2)$ and $P(3)$. Assume that $P(n)$ is true for some integer $n \geq 3$. To prove $P(n + 1)$, let us try to transfer a tower with $n + 1$ disks.

A critical moment occurs when the bottom disk is moving, from the first peg to the third. In order for this move to be possible, the n smaller disks must form a tower on the second peg. Thus before the move of the bottom disk, we must transfer a tower with n disks from the first peg to the second. Since $P(n)$ is true, this takes $2^n - 1$ moves. After the move of the bottom disk, we must complete the task by transferring the tower on the second peg, consisting of the n smaller disks, to the third peg. By $P(n)$, this also takes $2^n - 1$ moves. Together with the move of the bottom disk, the minimum number is $(2^n - 1) + 1 + (2^n - 1) = 2^{n+1} - 1$, which establishes $P(n + 1)$. Hence $P(n)$ is true for all integers $n \geq 1$.

We can express the solution to the problem of the Tower of Hanoi using a different terminology. Let a_n be the minimum number of moves to transfer a tower of height n from one peg to another. From our analysis above, we see that $a_n = 2a_{n-1} + 1$. This result, which defines a_n in terms of a_{n-1}, is an example of what is called a **recurrence relation**. Along with $a_1 = 1$, which is called an **initial value**, they define the sequence $\{a_n\}$ uniquely.

Using the recurrence relation and the initial value, we can generate additional terms of the sequence, as shown in the chart below.

$$a_1 = 1 \qquad a_2 = 3 \qquad a_3 = 7$$
$$a_4 = 15 \qquad a_5 = 31 \qquad a_6 = 63$$
$$a_7 = 127 \qquad a_8 = 255 \qquad a_9 = 511$$

We now consider a variant which we call the Twin Towers of Hanoi (see [1]). As before, there are three pegs in the playing board. There are n disks of sizes 1, 2, 3, ..., n. Those of odd sizes are stacked on the first peg, and those of even sizes are stacked on the third peg. On both pegs, the disks are in ascending order of size from the top. The rule is the same, in that we may only move a disk on top of a peg to the top of another peg, and a disk may not be placed on top of a smaller disk. Figure 5.2 illustrates the starting position of the case $n = 6$.

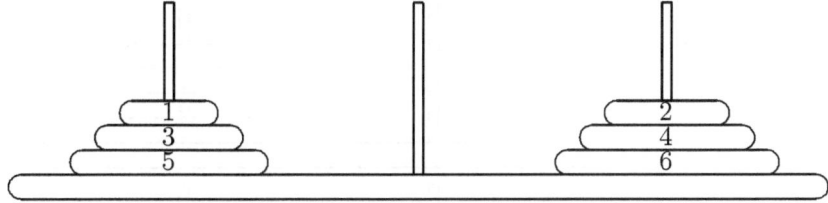

Figure 5.2

The objective is to have the two towers trade places. As in the Tower of Hanoi, a critical moment occurs when disk 6 is moving, from the third peg to the first. In order for this move to be possible, the 5 smaller disks must form a tower on the second peg, as illustrated in Figure 5.3. Thus before the move of disk 6, we must merge the disks 1 to 5 into a tower on the second peg.

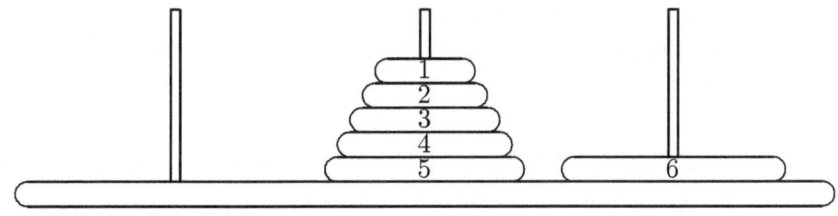

Figure 5.3

We have identified an intermediate objective for the Twin Towers of Hanoi, that of merging the n disks on the second peg. Figure 5.4 illustrates the case $n = 5$.

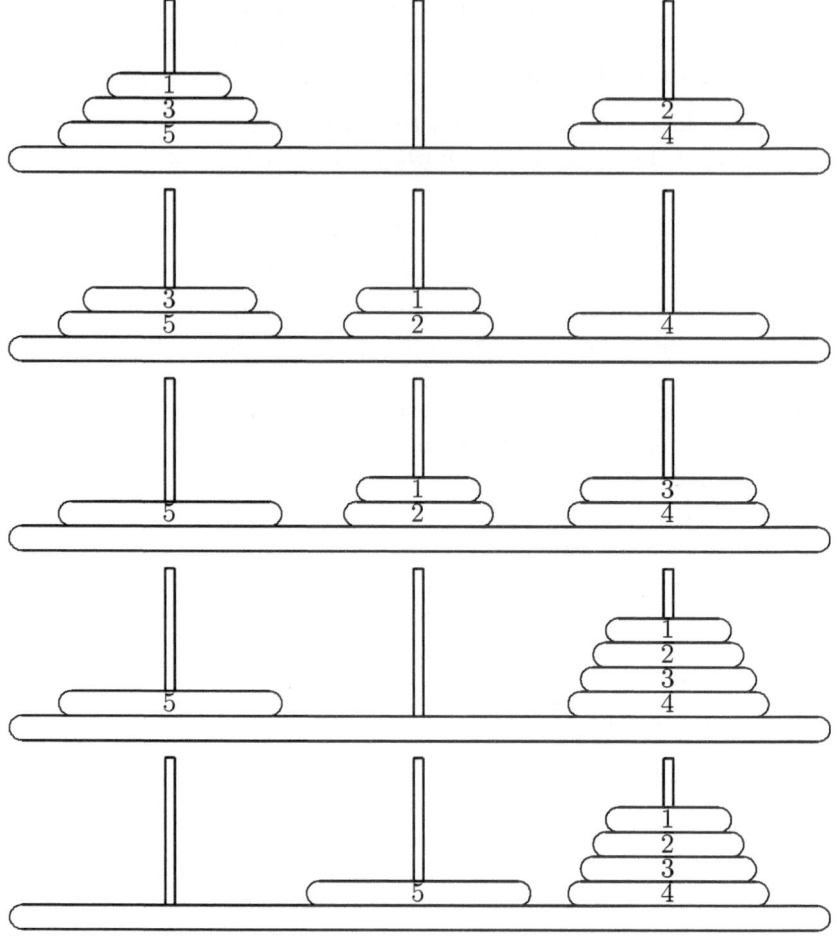

Figure 5.4

We first merge disks 1 and 2 on the second peg to pave the way for the move of disk 3. After disk 3 moves on top of disk 4, we transfer disks 1 and 2 on top of disk 3 to pave the way for the move of disk 5. After disk 5 moves to the second peg, we transfer disks 1, 2, 3 and 4 on top of it to complete the task.

In general, we merge disks 1, 2, 3, ..., $n - 3$ on the second peg, move disk $n - 2$ on top of disk $n - 1$, transfer disks 1, 2, 3, ..., $n - 3$ on top of disk $n - 2$, move disk n to the second peg and transfer disks 1, 2, 3, ..., $n - 1$ on top of disk n.

This means that if we let b_n be the minimum number of moves for merging n disks, then we have $b_n = b_{n-3} + 1 + a_{n-3} + 1 + a_{n-1}$. Since $a_n = 2^n - 1$, we have

$$b_n = b_{n-3} + 5 \cdot 2^{n-3}.$$

This is a three-step recurrence relation since b_n is not defined in terms of b_{n-1} but in terms of b_{n-3}. Hence we need three initial values. It is not hard to determine b_1, b_2 and b_3, and generate the chart of values below.

$$
\begin{array}{lll}
b_1 = 1 & b_2 = 2 & b_3 = 5 \\
b_4 = 11 & b_5 = 22 & b_6 = 45 \\
b_7 = 91 & b_8 = 182 & b_9 = 365
\end{array}
$$

We notice that b_n is either equal to $2b_{n-1}$ or $2b_{n-1} + 1$, but it is not immediately clear what the general formula for b_n is. We also notice that the numbers in the last column so far are all multiples of 5. Dividing them by 5 yields the quotients 1, 9 and 73. In the table of values of a_n given earlier, we notice that the numbers in the last column there are all multiples of 7. Dividing them by 7 yields the quotients 1, 9 and 73 also. Thus we suspect that b_n is roughly $\frac{5}{7}a_n$.

$$
\begin{array}{lll}
\frac{5}{7}a_1 = 1 - \frac{2}{7} & \frac{5}{7}a_2 = 2 + \frac{1}{7} & \frac{5}{7}a_3 = 5 \\
\frac{5}{7}a_4 = 11 - \frac{2}{7} & \frac{5}{7}a_5 = 22 + \frac{1}{7} & \frac{5}{7}a_6 = 45 \\
\frac{5}{7}a_7 = 91 - \frac{2}{7} & \frac{5}{7}a_8 = 182 + \frac{1}{7} & \frac{5}{7}a_9 = 365
\end{array}
$$

Thus we conjecture that b_n is equal to $\frac{5}{7}a_n$ rounded off to the nearest integer. More specifically, we conjecture that

$$
\begin{array}{lll}
b_n & = & \frac{5}{7}(2^n - 1) \quad \text{when } n \equiv 0 \pmod 3; \\
b_n & = & \frac{5}{7}(2^n - 2) + 1 \quad \text{when } n \equiv 1 \pmod 3 \text{ and} \\
b_n & = & \frac{5}{7}(2^n - 4) + 2 \quad \text{when } n \equiv 2 \pmod 3.
\end{array}
$$

We now use mathematical induction to prove these three formulas. First, let $n \equiv 0 \pmod 3$. We have $\frac{5}{7}(2^3 - 1) = 5 = b_3$. Suppose $b_n = \frac{5}{7}(2^n - 1)$ for some $n \geq 3$. Then

$$
\begin{aligned}
b_{n+3} & = b_n - 5 \cdot 2^n \\
& = \frac{5}{7}(2^n - 1) + 5 \cdot 2^n \\
& = \frac{5}{7}(2^n - 1 + 7 \cdot 2^n) \\
& = \frac{5}{7}(2^{n+3} - 1).
\end{aligned}
$$

The other two formulas can be proved in an analogous manner.

We now return to the main task of having the two towers trade places. Let the minimum number of moves required be c_n where n is the total number of disks. We have already identified that the critical moment is when disk n moves, either from the first peg to the third or vice versa. In order for this move to be possible, the $n-1$ smaller disks must form a tower on the second peg. This merger requires b_{n-1} moves. After the move of disk n, we must disperse the merged tower. This also takes b_{n-1} moves since the process is reversible. It follows that $c_n = 2b_{n-1} + 1$. We have the chart of values below.

$$c_1 = 1 \quad\quad c_2 = 3 \quad\quad c_3 = 5$$
$$c_4 = 11 \quad\quad c_5 = 23 \quad\quad c_6 = 45$$
$$c_7 = 91 \quad\quad c_8 = 183 \quad\quad c_9 = 365$$

It follows that c_n and b_n are identical, except for $n \equiv 2 \pmod 3$, when they differ by 1. It may be interesting to discover what may have caused this discrepancy.

Our solution of the recurrence relation $b_n = b_{n-3} + 5 \cdot 2^{n-3}$ largely depends on our observation that b_n is roughly $\frac{5}{7} a_n$. This we may obtain in a systematic manner. Because of the term $5 \cdot 2^{n-3}$, it is reasonable to conjecture that $b_n = K2^n$ for some constant K. Then $K2^n = K2^{n-3} + 5 \cdot 2^{n-3}$. Canceling the factor 2^{n-3} yields $8K = K + 5$. Hence $K = \frac{5}{7}$ so that $b_n = \frac{5}{7}2^n$.

This formula is a particular solution to our three-step recurrence relation. We point out that $b_n = \frac{5}{7}2^n + K_1$ for an arbitrary constant K_1 is also a solution, as shown below.

$$b_{n-3} + 5 \cdot 2^{n-3} = \frac{5}{7}2^{n-3} + K_1 + 5 \cdot 2^{n-3}$$
$$= \frac{5}{7}(2^{n-3} + 7 \cdot 2^{n-3}) + K_1$$
$$= \frac{5}{7}2^n + K_1$$
$$= b_n.$$

This is not surprising since $b_n = K_1$ is a solution to $b_n = b_{n-3}$. Since our specific solution depends on three initial values, we need a general solution with three arbitrary constants. So we need two more solutions of $b_n = b_{n-3}$. Let $b_n = x^n$ where $x > 0$. Then $x^n = x^{n-3}$, so that

$$0 = x^3 - 1 = (x - 1)(x^2 + x + 1).$$

The first factor yields the root $x = 1$ while the second factor yields complex roots ω and ω^2, where ω is a complex root of unity. It satisfies $\omega^3 = 1$ and $\omega^2 + \omega + 1 = 0$. The general solution is

$$b_n = \frac{5}{7}2^n + K_1 + K_2\omega^n + K_3\omega^{2n}.$$

We have

$$
\begin{aligned}
5 &= b_3 &&= \tfrac{40}{7} + K_1 + K_2 + K_3 \\
11 &= b_4 &&= \tfrac{80}{7} + K_1 + K_2\omega + K_3\omega^2 \\
22 &= b_5 &&= \tfrac{160}{7} + K_1 + K_2\omega^2 + K_3\omega
\end{aligned}
$$

It follows that

$$
\begin{aligned}
K_1 + K_2 + K_3 &= -\frac{5}{7} \\
K_1 + K_2\omega + K_3\omega^2 &= -\frac{3}{7} \\
K_1 + K_2\omega^2 + K_3\omega &= -\frac{6}{7}
\end{aligned}
$$

We can solve this system of three equations in three unknowns and obtain $K_1 = -\frac{2}{3}$, $K_2 = -\frac{2}{21} - \frac{1}{7}\omega$ and $K_3 = -\frac{2}{21} - \frac{1}{7}\omega^2$. However, we do not really need these values. For $n \equiv 0 \pmod 3$, we have

$$
\begin{aligned}
b_n &= \frac{5}{7}2^n + K_1 + K_2 + K_3 \\
&= \frac{5}{7}2^n - \frac{5}{7} \\
&= \frac{5}{7}(2^n - 1).
\end{aligned}
$$

For $n \equiv 1 \pmod 3$, we have

$$
\begin{aligned}
b_n &= \frac{5}{7}2^n + K_1 + K_2\omega + K_3\omega^2 \\
&= \frac{5}{7}2^n - \frac{3}{7} \\
&= \frac{5}{7}(2^n - 2) + 1.
\end{aligned}
$$

For $n \equiv 2 \pmod 3$, we have

$$
\begin{aligned}
b_n &= \frac{5}{7}2^n + K_1 + K_2\omega^2 + K_3\omega \\
&= \frac{5}{7}2^n - \frac{6}{7} \\
&= \frac{5}{7}(2^n - 4) + 2.
\end{aligned}
$$

Section 2. A Problem on Greatest Common Divisors

The point of mathematical induction is to establish that a certain pattern continues. However, it is necessary to recognize the pattern in the first place. We give here an example where the pattern is not all that easy to spot. Once we have it, the induction part is not that difficult.

We consider two companion sequences $\{a_n\}$ and $\{d_n\}$ defined by $a_1 = 3$, $a_n = a_{n-1} + d_{n-1}$ for $n \geq 2$ and d_n is the greatest common divisor of n and a_n for $n \geq 1$. It is only necessary to determine one of $\{a_n\}$ and $\{d_n\}$. We focus on $\{d_n\}$.

Let us generate some initial data.

n	a_n	d_n	n	a_n	d_n	n	a_n	d_n	n	a_n	d_n
1	3	1	31	78	1	61	162	1	91	192	1
2	4	2	32	79	1	62	163	1	92	193	1
3	6	3	33	80	1	63	164	1	93	194	1
4	9	1	34	81	1	64	165	1	94	195	1
5	10	5	35	82	1	65	166	1	95	196	1
6	15	3	36	83	1	66	167	1	96	197	1
7	18	1	37	84	1	67	168	1	97	198	1
8	19	1	38	85	1	68	169	1	98	199	1
9	20	1	39	86	1	69	170	1	99	200	1
10	21	1	40	87	1	70	171	1	100	201	1
11	22	11	41	88	1	71	172	1	101	202	101
12	33	3	42	89	1	72	173	1	102	303	3
13	36	1	43	90	1	73	174	1	103	306	1
14	37	1	44	91	1	74	175	1	104	307	1
15	38	1	45	92	1	75	176	1	105	308	7
16	39	1	46	93	1	76	177	1	106	315	1
17	40	1	47	94	47	77	178	1	107	316	1
18	41	1	48	141	3	78	179	1	108	317	1
19	42	1	49	144	1	79	180	1	109	318	1
20	43	1	50	145	5	80	181	1	110	319	11
21	44	1	51	150	3	81	182	1	111	330	3
22	45	1	52	153	1	82	183	1	112	333	1
23	46	23	53	154	1	83	184	1	113	334	1
24	69	3	54	155	1	84	185	1	114	335	1
25	72	1	55	156	1	85	186	1	115	336	1
26	73	1	56	157	1	86	187	1	116	337	1
27	74	1	57	158	1	87	188	1	117	338	13
28	75	1	58	159	1	88	189	1	118	351	1
29	76	1	59	160	1	89	190	1	119	352	1
30	77	1	60	161	1	90	191	1	120	353	1

It appears that $d_n = 1$ most of the time. When it is not, it is always a prime so far. Is this true? And if so, can we determine when the next prime will appear? The pattern so far is most obscure.

For $n \geq 3$, let us count the lengths of the blocks of successive 1s for d_n, separated by values of $d_n > 1$.

(0) 3	(1) 5	(0) 3	(4) 11	(0) 3	(10) 23	(0) 3	(22) 47	(0) 3
(1) 5	(0) 3	(49) 101	(0) 3	(2) 7	(4) 11	(0) 3	(7) 17	(0) 3

A pattern is emerging. The number of values of $d_n = 1$ preceding a value of $d_n = p$ for some prime p is equal to $\frac{p-3}{2}$ so far. Apparently, this pattern is not particularly useful since it is a mere observation after the fact, and does not indicate how the sequence $\{d_n\}$ will continue.

At the start, every other prime value of d_n is 3, and if we ignore them, the first four primes are 5, 11, 23 and 47. We have $5 \times 2 + 1 = 11$, $11 \times 2 + 1 = 23$ and $23 \times 2 + 1 = 47$. According to this pattern, the next value for $d_n > 1$, skipping over a 3, should be $2 \times 47 + 1 = 95$. However 95 is not a prime.

As it turns out, the next prime value of d_n is 5. Note that 5 is the smallest prime divisor of the composite number 95. An easy way out is to modify our pattern so that when a composite value for d_n is predicted, we take its smallest prime divisor instead.

Whether this is correct remains to be seen, but in any case, now that we are back to 5, how are we going to predict when the next prime will appear? Moreover, the pattern of every other prime value of d_n being 3 does not hold out either.

We now reexamine our initial data, focusing on a_n. Another pattern emerges: $a_{n+1} = 3n$ if and only if d_n is prime.

At this point, we digress and introduce a symbol for the greatest common divisor of two positive integers a and b. The usual notation $\gcd(a, b)$ is cumbersome, while the simplified notation (a, b) can be ambiguous. We want a notation which emphasizes the fact that finding the greatest common divisor of two numbers is a binary operation, just as finding their sum or their product.

Let a and b be positive integers. We use $a \bigtriangleup b$ to denote their greatest common divisor, and $a \bigtriangledown b$ to denote their least common multiple. They mirror the logical connectives $\vee$ and $\wedge$ which underlie their definitions. With the new symbols, statements about these concepts become very succinct. For instance, the key equation in the Euclidean Algorithm is $a \bigtriangleup b = a \bigtriangleup (b - a)$.

We now return to the sequences. Suppose we have $a_{n+1} = 3n$. We want to know for what range of k will we have $a_{n+1+k} = 3(n)+k$. This is certainly true for $k = 0$. If $d_{n+1} = 1$, then $a_{n+2} = a_{n+1} + 1 = 3(n) + 2$, and it is true for $k = 1$. This will be true for $k + 1$ as long as $d_{n+k} = 1$. During this run, $a_{n+1+k} - (n+1+k) = 2n - 1$. Hence d_{n+1+k} is a divisor of $2n - 1$. It follows that the next value of $d_n > 1$ occurs when it is equal to the smallest prime divisor of $2n - 1$.

Suppose that $a_{n+1} = 3n$, and the smallest prime divisor of $2n - 1$ is p. We will have $a_{n+1+k} = 3n + k$ for $0 \le k \le \frac{p-3}{2}$. We now prove this by induction. Clearly, this holds for $k = 0$. Suppose $a_{n+1+k} = 3n + k$ for some $k < \frac{p-3}{2}$. Then

$$
\begin{aligned}
d_{n+1+k} &= a_{n+1+k} \triangle (n + 1 + k) \\
&= (3n + k) \triangle (n + 1 + k) \\
&= (2n - 1) \triangle (2n + 2 + 2k) \\
&= (2n - 1) \triangle (2k + 3) \\
&= 1,
\end{aligned}
$$

because $2k + 3 < 2\frac{p-3}{2} + 3 = p$. Hence $a_{n+1+(k+1)} = 3n + (k + 1)$. This completes the inductive argument. Thus $d_{n+1} = d_{n+2} = \cdots = d_{n+\frac{p-3}{2}} = 1$ and $a_{n+\frac{p-3}{2}} = 3n + \frac{p-3}{2}$. Now

$$
\begin{aligned}
d_{n+1+\frac{p-3}{2}} &= a_{n+1+\frac{p-3}{2}} \triangle (n + 1 + \frac{p-3}{2}) \\
&= 3(n + \frac{p-3}{2}) \triangle (n + 1 + \frac{p-3}{2}) \\
&= (2n - 1) \triangle (2(n+1) + p - 3) \\
&= (2n - 1) \triangle p \\
&= p.
\end{aligned}
$$

It follows that $a_{n+1+\frac{p-3}{2}} = 3(n + \frac{p-3}{2}) + p = 3(n + \frac{p-1}{2})$.

This validates the earlier observation that $a_{n+1} = 3n$ if and only if d_n is prime. For any such value of n, the next such value is $n + \Delta$, where $\Delta = \frac{p-1}{2}$ and p is the smallest prime divisor of $m = 2n - 1$. We can now summarize the construction of all such values of n as follows.

n	m	p	Δ	n	m	p	Δ	n	m	p	Δ
2	3	3	1	23	45	3	1	101	201	3	1
3	5	5	2	24	47	47	23	102	203	7	3
5	9	3	1	47	93	3	1	105	209	11	5
6	11	11	5	48	95	5	2	110	219	3	1
11	21	3	1	50	99	3	1	111	221	13	6
12	23	23	11	51	101	101	50	117	233	233	105

The next such value of n will be 117+105=222. This established pattern determines both $\{a_n\}$ and $\{d_n\}$.

Having dealt successfully with the problem when $a_1 = 3$, it is natural to ask what the situation is for other values of a_1. Here the pattern is even more obscure.

Again, we generate some initial data.

n	a_n	d_n	a_n	d_n	a_n	d_n	a_n	d_n	a_n	d_n	a_n	d_n
1	1	1	2	1	3	1	4	1	5	1	6	1
2	2	2	3	1	4	2	5	1	6	2	7	1
3	4	1	4	1	6	3	6	3	8	1	8	1
4	5	1	5	1	9	1	9	1	9	1	9	1

Within the first four rows, the sequences generated by $a_1 = 2$ coincide with those generated by $a_1 = 1$, and the sequences generated by $a_1 = 4$, 5 or 6 coincide with those generated by $a_1 = 3$. It follows that these coincidences will hold if we add $6k$ to a_1 for any positive integer k. Hence the only meaningful values for a_1 are numbers congruent modulo 6 to 1 or 3.

From our analysis of the case $a_1 = 3$, the key is finding a value of n for which $a_{n+1} = 3n$. For $a_1 = 7$, $d_2 = 2$ is prime but $a_3 \neq 3(2)$. On the other hand, $a_5 = 3(4)$, but d_4 is not prime. However, this value correctly predicts the next value $a_8 = 3(7)$, and this time, $d_7 = 7$ is prime. After these early anomalies, the sequences behave themselves. For $a_1 = 9$, the sequences $\{a_n\}$ and $\{d_n\}$ behave themselves right from the beginning.

It would be tempting to conjecture that d_n is never composite for any value of a_1. However, programming wizard **Dr. George Sicherman** of New Jersey discovers that $d_{18} = 9$ for $a_1 = 531$. Thus the general problem remains open.

Section 3. Congo Bongo

The following problem was posed in the Senior A-Level paper in the 2009 Fall Round of the International Mathematics Tournament of the Towns. It was solved on the spot by Hsin-Po Wang, a member of Chiu Chang Mathematical Circle. His solution was published in [6].

An expedition into Congo uncovered a round treasure chest. Evenly spaced around the circumference were n identical bongo drums. A scroll attached to the chest said that there was a monkey inside each bongo drum. The monkey might be standing upright or doing a handstand, but it was not visible from outside. One might hit a number of bongo drums at the same time. If a bongo drum was hit, the monkey inside would change its posture, from right side up to upside down or vice versa. The treasure chest would open if and only if all monkeys were right side up, or all were upside down. However, the treasure chest would spin on its vertical axis each time some bongo drums were hit, so that in general it would not be possible to tell which had just been hit. For what values of n could the treasure chest be opened for sure?

The case $n = 1$ is trivial since the chest will open automatically. The case $n = 2$ is not hard either. If the chest is not already open, hitting either of the drums will work. For the purpose of later reference, we call this operation **A**.

For $n = 3$ we hit only one drum. This is because hitting two is just the same as hitting the third one, and hitting all three serves no purpose. Suppose the chest is not already open. This means that two monkeys have the same posture while the third one has the opposite posture. If we hit either of the first two drums, we are left with essentially the same situation as before, and this can continue indefinitely. Hence we cannot be sure to open the chest.

Suppose n is not a power of 2. Then it has a smallest odd divisor $d > 1$. Focus on d evenly spaced drums and ignore all others. Initially, the monkeys in these drums are not all in the same posture. Since d is odd, the number of drums with right side up monkeys inside is different from the number of drums with upside down monkeys inside at any time. In order for us to succeed, we must at some point hit all bongo drums in one group. However, it is always possible for us to hit at least one drum from the opposite group, and we can be kept from opening the treasure chest forever.

So the task is only possible if n is a power of 2, but is this so for all powers of 2? The next case is $n = 4$, and we have to consider sixteen possible configurations regarding the postures of the monkeys.

Let 0 or 1 indicate whether a monkey is right side up or upside down. If we are really lucky, the initial state may already be (0,0,0,0) or (1,1,1,1). We may regard them as the same state, and call it State 0. It is also referred to as an *absorbing* state, in that once we enter it, we do not leave, since the chest will have opened already.

Note that we will be hitting one or two drums, and in the latter case, we may be hitting two adjacent drums or two opposite drums. Let us examine which operation may take a particular configuration to State 0.

Suppose the configuration is (0,1,0,1) or (1,0,1,0). The chest will open if we hit two opposite drums. We call this operation **A**, and put these two configurations in State 1.

Suppose the configuration is (1,1,0,0), (1,0,0,1), (0,0,1,1) or (0,1,1,0). The chest may open if we hit two adjacent drums. We call this operation **B**, and put these four configurations in State 2. Note that if we do not get to State 0, we will get to State 1. On the other hand, if we perform operation A while we are in State 2, we will stay in State 2.

Suppose the configuration is any of the remaining eight, namely (1,0,0,0), (0,1,0,0), (0,0,1,0), (0,0,0,1), (0,1,1,1), (1,0,1,1), (1,1,0,1) or (1,1,1,0). The chest may open if we hit just one drum. We call this operation **C**, and put these eight configurations into State 3. Note that if we do not get to State 0, we will get to either State 1 or State 2. On the other hand, if we perform operation A or B while we are in State 3, we stay in State 3.

The state transition diagram is shown in Figure 5.5.

State 0   State 1   State 2   State 3

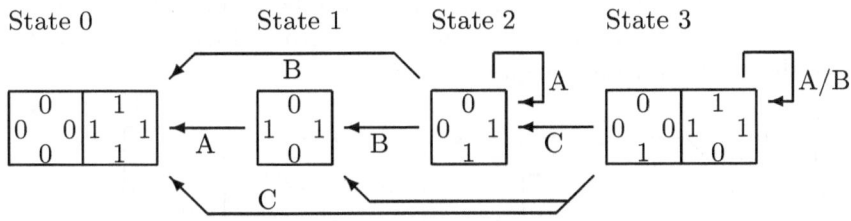

Figure 5.5

Remember that we cannot see the monkeys, so that we cannot tell in which state we are, unless we are already in State 0. Let us see what happens if we perform operation A. If we are in State 1, we will get to State 0 and know it. If we are in State 2 or State 3, nothing changes. So we can only conclude that we cannot be in State 1 just after performing operation A.

What happens if we perform operation B right after performing operation A. We know that we are not in State 0 or 1. If we are in State 2, we will either get to State 0 or State 1. If we are in State 3, nothing changes. So all we know is that we will not be in State 2 just after performing operation B.

We must not hurry and perform operation C. This is because while State 1 has been cleared out by operation A, it may be restocked by operation B. Thus we should perform operation A once more to make sure State 1 is cleared out as well as State 2.

We are now ready to perform operation C, which will clear State 3. We must then clear States 1 and 2 again, and this can be accomplished by performing operation A, operation B and operation A again in that order. Thus the chest will open in seven moves: **ABACABA**. Note that this is exactly the same solution as the Tower of Hanoi with three disks at the beginning of Section 1.

The solution to the Congo Bongo problem with four drums is already reasonably complex. It would be hard to visualize what a solution to the eight-drum case may entail, let alone the general case. Thus we must use a more systematic approach. The Principle of Mathematical Induction comes immediately to mind as we are trying to prove for all positive integers n the statement $P(n)$, which states that the chest will open whenever the number of drums is 2^n.

In order to reduce the case with 2^{n+1} drums to the case with 2^n drums, we use the strategy in Section 2. We will identify pairs of drums as a single drum, just as we combine two slices of bread into a sandwich.

How can we choose partners for the drums? We must pair a drum with the one directly opposite. This relation is not affected by the spinning of the chest, whereas partnership under any other relation becomes ambiguous.

Let us consider $P(3)$. The simplest states are those in which the monkeys in each opposite pair of drums have the same posture. Figure 5.6 shows this part of the state transition diagram.

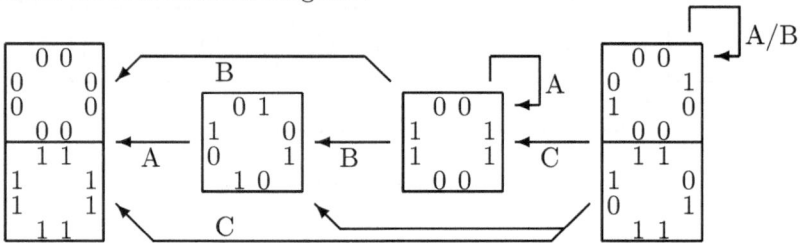

Figure 5.6

It is really the same as Figure 5.5, except that operation A means hitting every other drum; operation B means hitting any two adjacent pairs of opposite drums; and operation C means hitting any pair of opposite drums. By performing the sequence **ABACABA**, the treasure chest will open.

These states together form an expanded absorbing state in the overall diagram in Figure 5.7. It is the box marked 4. For $0 \leq m \leq 4$, the box marked m contains all states with m opposite pairs of drums containing monkeys with the same posture. It looks suspiciously like Figure 5.5, except that States 4 and 0 are no longer equivalent. The states with 2 matching pairs are classified according to whether they are alternating or adjacent. The former states are grouped under State 2 while the latter states are grouped under State 2$'$.

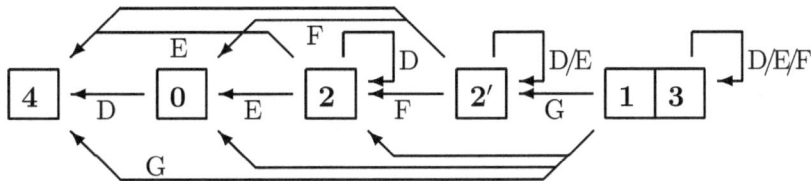

Figure 5.7

Operation **D** means hitting any 4 adjacent drums. Operation **E** means hitting any two drums separated by one other drum. Operation **F** means hitting any two adjacent drums. Operation **G** means hitting any drum. If we denote the sequence ABACABA by **X**, then the sequence which will open the chest is

XDXEXDXFXDXEXDXGXDXEXDXFXDXEXDX.

We keep repeating X so that whenever we enter State 4, we will not return to another State. Whatever state the chest is in initially, it will open by the end of this sequence.

This case is isomorphic to the Tower of Hanoi with seven disks.

There is a related problem posed by Martin Gardner [2] in his famous *Mathematical Games* column in the magazine **Scientific American**. That version has been studied by others. See [3], [4] and [5].

Exercises

1. An arbitrary square of a $2^n \times 2^n$ chessboard is removed. Prove that
   the remaining part can be covered by copies of the shape in Figure
   5.8. When placed on the chessboard, it covers exactly three of its
   squares. The copies may not overlap or stick out beyond the border
   of the chessboard.

Figure 5.8

2. In the problem on greatest common divisors, suppose that
   (a) $a_1 = 13$; (b) $a_1 = 15$; (c) $a_1 = 19$; (d) $a_1 = 21$.
   Find the first value of n for which $a_{n+1} = 3n$ and d_n is prime, and
   predict when the next such value of n will occur.

3. An expedition into Tonga uncovered a round treasure chest. Evenly
   spaced around the circumference were n identical conga drums. A
   scroll attached to the chest said that there was a monkey inside each
   conga drum. The monkey might be standing upright or doing a handstand, but it was not visible from outside. One might touch two conga
   drums at the same time, and the monkeys inside became visible. One
   might then decide to hit neither, either or both conga drums. If a conga
   drum was hit, the monkey inside would change its posture, from right
   side up to upside down or vice versa. The treasure chest would open
   if and only if all monkeys were right side up, or all were upside down.
   However, the treasure chest would spin on its vertical axis each time
   some conga drums were hit, so that in general it would not be possible to tell which had just been hit. Moreover, the monkeys became
   invisible from outside again. How could the treasure chest be opened
   for sure if

   (a) $n = 3$;
   (b) $n = 4$?

Bibliography

[1] Jack Chen, Richard Mah and Steven Xia, The twin towers of Hanoi, Mathematics Competitions **26** (2013) #1, 59–68.

[2] Martin Gardner, The rotating table, in "Fractal Music, Hypercards and more Mathematical Recreations", W. H. Freeman & Co., (1992) 107–108 and 181–184.

[3] W. Laaser and L. Ramshaw, Probing the rotating table, in "The Mathematical Gardner", edited by David Klarner, Wadsworth, (1981) 288–307.

[4] Ted Lewis and Steve Willard, The rotating table, *Mathematics Magazine*, **53** (1980) 174–179.

[5] A. Stanger, Variations on the rotating table problem, *Journal of Recreational Mathematics*, **19** (1987) 307–308 and **20** (1988) 312–314.

[6] Hsin-Po Wang, Congo Bongo, *Math Horizons*, September (2010) of 18–21.

Chapter Six: Number Triangles

Section 1. Pascal's Triangle

Imagine that you are the owner of a small coffee shop, and you have just imported a box of the finest Columbian coffee beans. As you open it, you savor the aroma. Suddenly, your smile turns into a frown as you realize that some of the essence of the coffee has evaporated into thin air.

We use the following **mathematical model** to measure the loss. We assume that there are n kilograms of coffee beans initially, where n is a positive integer, and that you will use 1 kilogram each day. Each kilogram in a box loses 1 aroma point every time the box is opened. Fortunately, you have some empty boxes which help in reducing future losses. Let k be the number of boxes available, including the one in which the coffee beans come. You want to minimize the total number of points lost.

Let us first work out an example with $k = 2$ and $n = 6$. After checking all cases, we find this optimal strategy. Let the boxes be numbered 1 and 2.

Day Number	Open Box	Points Lost	Shift to Box 2	Amount in Box 1	Amount in Box 2
1	1	6	2 kg	3 kg	2 kg
2	2	2		3 kg	1 kg
3	2	1		3 kg	
4	1	3	1 kg	1 kg	1 kg
5	2	1		1 kg	
6	1	1			
Total =		14			

We now consider the general problem. Clearly, counting the number of points lost each day is not a promising approach, especially since we do not even know how many kilograms of coffee beans are to be transferred from which box to which, and when. The main idea behind our attack of this problem is to count the number of points lost by each **kilogram**.

The number of points each kilogram of coffee beans loses is equal to the number of times it is exposed. We keep track of this by putting a label on each kilogram. Number the boxes 1 to k. A label is initially empty. Every time the kilogram is exposed while in box i, add an i to the end of its current label. The label changes progressively until the kilogram is used up. Its length at that time is the total number of points lost.

Each label starts with a 1. By symmetry, we can arrange to have no more coffee beans in a box with a higher number than in a box with a lower one. Each day, we always open the non-empty box with the highest number.

© Springer International Publishing AG 2018
A. Liu, *S.M.A.R.T. Circle Projects*, Springer Texts in Education, DOI 10.1007/978-3-319-56811-9_6

Thus we never transfer coffee beans from a box with a higher number to a box with a lower one. This means that the terms in each label are non-descending. Since exactly one kilogram of coffee beans is used each day, no two kilograms can have the same label. What we want is a set of the shortest n labels.

Let us return to our example with $k = 2$ and $n = 6$. There is only one label of length 1, namely 1. There are two labels of length 2 and three labels of length 3. They are 11, 12, 111, 112 and 122. Thus the minimum number of points lost is $1 + 2 + 2 + 3 + 3 + 3 = 14$. This justifies that our strategy is indeed optimal. In fact, it is the only one that leads to the optimal result, since the labels tell us precisely what to do.

Each kilogram is exposed in box 1 on day 1. The kilogram labeled 1 is used immediately. The kilograms labeled 12 and 122 must be shifted to box 2 then. They are used on days 2 and 3. The remaining three kilograms are all exposed in box 1 on day 4. The kilogram labeled 11 is used immediately, while the kilogram labeled 112 must be shifted to box 2. It is used on day 5, while the kilogram labeled 111 stays in box 1 throughout, and is used on day 6.

The general problem is solved if we can count the number of distinct labels of length ℓ with non-descending terms such that the first is 1 and none exceeds k. As another example, consider the case $k = 3$ and $\ell = 5$. There are 15 such labels, listed below.

11111	11133	12222
11112	11222	12223
11113	11223	12233
11122	11233	12333
11123	11333	13333

Counting the labels directly is no easy matter either. We now change each into a binary sequence as follows. Write down a number of 0s equal to the number of 1s in the label. Insert a 1 after this block. Then write down a number of 0s equal to the number of 2s, followed by another 1, and so on. Note that each binary sequence consists of k 1s and ℓ 0s, starts with a 0 and ends with a 1.

As an example, consider the label 11122. We start off with three 0s followed by a 1. Then we write down two 0s followed by a 1. Finally, since the label contains no 3s, we just write down one more 1, yielding the binary sequence 00010011. Conversely, consider the binary sequence 01000101. We start off with one 1, followed by three 2s and then one 3, yielding the label 12223.

It is clear that each label is matched with a unique binary sequence whose first term is 0 and last term 1, and vice versa. The corresponding binary sequences are listed after the labels in the chart below.

11111	00000111	11133	00011001	12222	01000011
11112	00001011	11222	00100011	12223	01000101
11113	00001101	11223	00100101	12233	01001001
11122	00010011	11233	00101001	12333	01010001
11123	00010101	11333	00110001	13333	01100001

It is not too difficult to count such binary sequences. As noted before, they are of length $k + \ell$. Since the first term is always 0 and the last term is always 1, we only need to consider the $k + \ell - 2$ terms in between. They consist of $\ell - 1$ 0s and $k - 1$ 1s, and all we have to do is count the number of ways of placing the 1s, by choosing $k - 1$ of the $k + \ell - 2$ available positions.

This is a very basic problem. A **combination** is a selection of some of the objects in a set. The number of ways of choosing k objects from a set of n objects is denoted by $\binom{n}{k}$, which is verbalized as "n choose k". We begin by making a few simple observations:

- $\binom{n}{k} = 0$ if $n < k$;

- $\binom{n}{0} = 1$;

- $\binom{n}{n} = 1$;

- $\binom{n}{k} = \binom{n}{n-k}$.

In the last item, the number of ways of choosing k objects from a set of n is equal to the number of ways of eliminating $n - k$ objects from the set of n. This is a prototype of a *combinatorial* argument, which we will pursue further. Such an argument is also featured in the proof of the following result.

Pascal's Formula: $\binom{n}{k} = \binom{n-1}{k-1} + \binom{n-1}{k}$.

Proof:
Fix an arbitrary one of the n objects. If we take it, we can choose $k - 1$ more from the remaining $n - 1$. If we leave it, we will be choosing all k from the remaining $n - 1$. The desired result follows from the Addition Principle since we either take it or leave it.

We can compute the value of $\binom{n}{k}$ recursively by building the famous Pascal's Triangle, the first few rows of which are shown in Figure 6.1.

$$\binom{0}{0}=1$$

$$\binom{1}{0}=1 \qquad\qquad \binom{1}{1}=1$$

$$\binom{2}{0}=1 \qquad\qquad \binom{2}{1}=2 \qquad\qquad \binom{2}{2}=1$$

$$\binom{3}{0}=1 \qquad \binom{3}{1}=3 \qquad\qquad \binom{3}{2}=3 \qquad\qquad \binom{3}{3}=1$$

$$\binom{4}{0}=1 \qquad \binom{4}{1}=4 \qquad\qquad \binom{4}{2}=6 \qquad\qquad \binom{4}{3}=4 \qquad\qquad \binom{4}{4}=1$$

Figure 6.1

It is also possible to express $\binom{n}{k}$ directly in terms of the **factorial** function, For a positive integer n, we define $n! = n(n-1)(n-2)\cdots 3\cdot 2\cdot 1$. We also take $0!=1$.

Factorial Formula.
For integers $n \geq k \geq 0$, $\binom{n}{k} = \frac{n!}{k!(n-k)!}$.

Proof:
We first verify the boundary conditions on Pascal's Triangle. We have $\frac{n!}{0!(n-0)!} = 1 = \binom{n}{0}$ and $\frac{n!}{n!(n-n)!} = 1 = \binom{n}{n}$. Thus the first two rows of Pascal's Triangles conform. We now proceed row by row. In each, the two outside entries have been verified. Each inside entry is the sum of two entries of the preceding row by Pascal's Formula. Indeed,

$$
\begin{aligned}
\binom{n}{k} &= \binom{n-1}{k-1} + \binom{n-1}{k} \\
&= \frac{(n-1)!}{(k-1)!(n-k)!} + \frac{(n-1)!}{k!(n-k-1)!} \\
&= \frac{(n-1)!}{k!(n-k)!}(k+(n-k)) \\
&= \frac{n!}{k!(n-k)!}.
\end{aligned}
$$

Returning to the original problem (see [2]), the number of binary sequences consisting of $\ell-1$ 0s and $k-1$ 1s is $\binom{k+\ell-2}{k-1}$. When $k = 3$ and $\ell = 5$, $\binom{k+\ell-2}{k-1} = \binom{6}{2} = 15$. Hence there are indeed 15 labels of length five, as we have seen earlier.

For n kilograms of coffee beans, let the longest labels have length m. This means that we use all labels of length less than m, and as many labels of length m as needed to bring the total up to n. Hence m is the largest integer such that the total number N of labels of length from 1 to $m-1$ is less than n. Clearly,

$$
N = \binom{k-1}{k-1} + \binom{k}{k-1} + \binom{k+1}{k-1} + \cdots + \binom{k+m-3}{k-1}.
$$

Using $\binom{k-1}{k-1} = 1 = \binom{k}{k}$ and Pascal's Formula, we can simplify this expression as follows.

$$
\begin{aligned}
N &= \binom{k-1}{k-1} + \binom{k}{k-1} + \binom{k+1}{k-1} + \cdots + \binom{k+m-3}{k-1} \\
&= \binom{k}{k} + \binom{k}{k-1} + \binom{k+1}{k-1} + \cdots + \binom{k+m-3}{k-1} \\
&= \binom{k+1}{k} + \binom{k+1}{k-1} + \cdots + \binom{k+m-3}{k-1} \\
&= \binom{k+2}{k} + \cdots + \binom{k+m-3}{k-1} \\
&= \cdots \\
&= \binom{k+m-3}{k} + \binom{k+m-3}{k-1} \\
&= \binom{k+m-2}{k}.
\end{aligned}
$$

For any positive integer n, let m be the largest positive integer such that $n > \binom{k+m-2}{k}$. Let $r = n - \binom{k+m-2}{k}$, where

$$
1 \le r \le \binom{k+m-1}{k} - \binom{k+m-2}{k} = \binom{k+m-2}{k-1}.
$$

Then the n labels consists of $\binom{k-1}{k-1}$ of length 1, $\binom{k}{k-1}$ of length 2, ..., $\binom{k+m-3}{k-1}$ of length $m-1$, and r of length m. It follows that the minimum number of points lost is

$$
\binom{k-1}{k-1} 1 + \binom{k}{k-1} 2 + \cdots + \binom{k+m-3}{k-1}(m-1) + rm,
$$

and that this optimal value can be attained.

In our original example, $n = 6$ and $k = 2$. Now m is determined by $6 > \binom{m}{2}$, so that $m = 3$. Hence $r = 6 - \binom{3}{2} = 3$ and the minimum number of points lost is $\binom{1}{1}1 + \binom{2}{1}2 + \binom{3}{1}3 = 14$, as we found. If $k = 3$, then $m = 3$, $r = 2$ and 13 points lost. If $k = 4$, then $m = 3$, $r = 1$ and 12 points lost. If $k = 5$, then $m = 2$, $r = 5$ and 11 points lost. This is the best that can be done with $n = 6$, and we leave to the reader the details of how to move the kilograms of coffee around.

This was the work [2] of Circle members Robert Barrington Leigh and Richard Ng.

Section 2. Rascal's Triangle

An I.Q. test question asked for the next row of numbers in the triangular array in Figure 6.2.

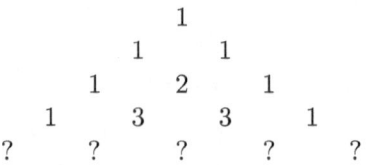

$$\begin{array}{ccccccccc}
& & & & 1 & & & & \\
& & & 1 & & 1 & & & \\
& & 1 & & 2 & & 1 & & \\
& 1 & & 3 & & 3 & & 1 & \\
? & & ? & & ? & & ? & & ?
\end{array}$$

Figure 6.2

Our answer was "1 4 5 4 1".

"WRONG!" said our teacher. "The answer is 1 4 6 4 1, part of the well-known Pascal triangle."

"The question does not ask for the next row in the Pascal triangle," we complained. "Without telling us what the triangle is, any five numbers should be acceptable."

"Perhaps it was not stated clearly," our teacher conceded, "but the purpose of the question is to find a way of filling in the new row according to some simple rule. For the Pascal triangle, the rule is that the outside numbers on each row are 1s and the inside numbers are determined by the *inverted triangle formula: South=West+East*. For example, 6=3+3 (see Figure 6.3)."

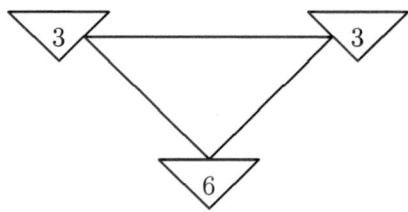

Figure 6.3

"Our rule is that the outside numbers on each row are 1s and the inside numbers are determined by the *diamond formula:*

$$South = (West \times East + 1) \div North.$$

For example, $5 = (3 \times 3 + 1) \div 2$ (see Figure 6.4)."

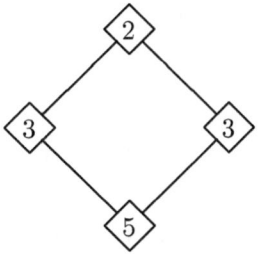

Figure 6.4

"That is not as simple as the inverted triangle formula," said our teacher. "Moreover, the diamond formula involves division. How do you even know that all numbers in your triangle are integers?"

"We have worked out the next few rows." We produced a piece of paper containing Figure 6.5. "See!"

```
                              1
                        1           1
                  1           2           1
            1           3           3           1
      1           4           5           4           1
1           5           7           7           5           1
1     6           9          10           9           6     1
1     7     11          13          13          11     7     1
```

Figure 6.5

"That may be, but you do not know for sure. The first non-integer may just be round the corner. In the Pascal triangle, only addition is involved, and we can be sure that all numbers are integers."

We went away, still believing we were right. However, before showing that our triangle was simpler than the Pascal triangle, we had to show that all numbers in our triangle were indeed integers. How?

Since our triangle is not quite the Pascal triangle, we call it the *rascal triangle*. We conceive of the two triangles as finished products rather than as expanding structures. We notice that the two triangles are the same for the first two diagonals running from northwest to southeast. The 0th diagonal consists of all 1s while the 1st diagonal consists of the positive integers.

The 2nd diagonal in the Pascal triangle is 1, 3, 6, 10, 15, ... Each term after the first is obtained from the preceding one by adding successively larger integers, namely, 1+2=3, 3+3=6, 6+4=10, 10+5=15, ... The second diagonal in the rascal triangle is 1, 3, 5, 7, 9, ..., which consists of just the odd numbers. Surely, we have a simpler pattern, namely, 1+2=3, 3+2=5, 5+2=7, 7+2=9, ...

The third diagonal in the Pascal triangle is 1, 4, 10, 20, 35, ..., and it is already not quite easy for us to see a pattern. The third diagonal in the rascal triangle is 1, 4, 7, 10, 13, ... Each term is obtained from the preceding one by simply adding 3. In fact, the mth diagonal in the rascal triangle starts with 1 as its 0-th term, and each subsequent term is obtained from the preceding one by simply adding m. Hence the nth term on this diagonal is $mn + 1$.

If this pattern continues, all numbers in the rascal triangle will be integers. To show this, consider the triangle in which this pattern does continue. All we have to do is show that it is the same as the rascal triangle. In other words, the diamond formula holds here. So suppose North is the nth term on the mth diagonal. Then West is the nth term on the $(m + 1)$st diagonal, East is the $(n + 1)$st term on the mth diagonal, and South is the $(n + 1)$st term on the $(m + 1)$st diagonal. The calculation below shows that the diamond formula holds, so that all numbers in the rascal triangle are indeed integers!

$$
\begin{aligned}
& \frac{West \times East + 1}{North} \\
=\ & \frac{(m(n + 1) + 1)((m + 1)n) + 1) + 1}{mn + 1} \\
=\ & \frac{(mn + m + 1)(mn + n + 1) + 1}{mn + 1} \\
=\ & \frac{m^2n^2 + m^2n + mn^2 + 3mn + m + n + 2}{mn + 1} \\
=\ & \frac{mn(mn + 1) + m(mn + 1) + n(mn + 1) + 2(mn + 1)}{mn + 1} \\
=\ & mn + m + n + 2 \\
=\ & (m + 1)(n + 1) + 1 \\
=\ & South.
\end{aligned}
$$

The k-th number on the r-th row of the Pascal triangle is $\binom{r}{k} = \frac{r!}{k!(r-k)!}$. When written in full, it is

$$
\frac{r(r - 1)(r - 2) \cdots 3 \cdot 2 \cdot 1}{k(k - 1)(k - 2) \cdots 3 \cdot 2 \cdot 1 \cdot (r - k)(r - k - 1)(r - k - 2) \cdots 3 \cdot 2 \cdot 1},
$$

which has multiplications and divisions galore.

In the rascal triangle, the kth number on the rth row is the kth number on the $(r-k)$th diagonal. Hence this number is $k(r-k)+1$. Which triangle is simpler now?

The three little rascals, Circle member **Angus Tulloch** and his international friends Alif Anggoro from Indonesia and Eddy Liu from the United States, gained quite a bit of notoriety when this work [1] was published.

Section 3. Triangles of Absolute Difference

In 1976, George Sicherman thought of the following problem while watching a pool game in Buffalo. Could the fifteen pool balls be arranged in the usual triangular array so that apart from the back row of five numbers, each number in a subsequent row is the absolute difference of the two numbers immediately behind it?

As an illustration, possible arrangements with the pool balls numbered from 1 to 6 and from 1 to 10 are shown in Figure 6.6.

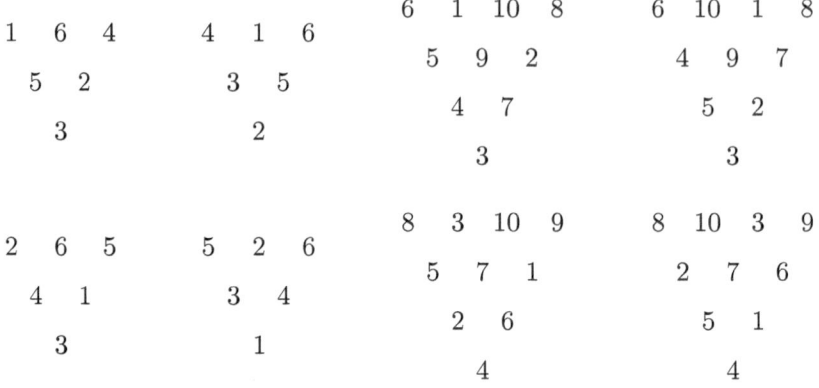

Figure 6.6

George found a solution to his problem, and established its uniqueness up to reflection about the triangle's vertical axis of symmetry. He communicated the problem to Martin Gardner [5] who published the problem in his famed column *Mathematical Games* in the magazine *Scientific American* in April 1977 [4].

A **triangle of absolute difference**, or *TOAD* for short, is defined to be a triangular array of integers having the following properties.

1. There are n integers on the top row for some positive integer n.

2. Each row below has one less number than the row above it.

3. Each number below the top row is the absolute difference of the two adjacent numbers in the row immediately above.

4. The integers are from 1 to $1 + 2 + \cdots + n = \frac{n(n+1)}{2}$, each appearing exactly once.

Such a TOAD is said to be of order n. We shall prove that TOADs of order greater than 5 do not exist.

Note that the reflection of a TOAD about its vertical axis of symmetry is another TOAD. We will treat each pair of mirror twins as a single TOAD. Without loss of generality, we may use either form of the TOAD at our convenience.

We now study the anatomy of a TOAD. It has a *spine* defined by the following construction. We call the bottom number of a TOAD its *foot*. From the foot, we draw two line segments connecting it to the two numbers of which it is the absolute difference, a thin line segment to the smaller one and a thick line segment to the larger one. The smaller number is called a *hand* and no further line segments are drawn from it.

From the larger number, we draw two more line segments as before. This is continued until the spine reaches the top row. The larger number on the top row is called its *head*. This is illustrated in Figure 6.7 for the unique TOAD of order 5. The spine consists of the numbers 5 (foot), 9, 11, 12 and 15 (head) while the hands consist of the numbers 4, 2, 1 and 3.

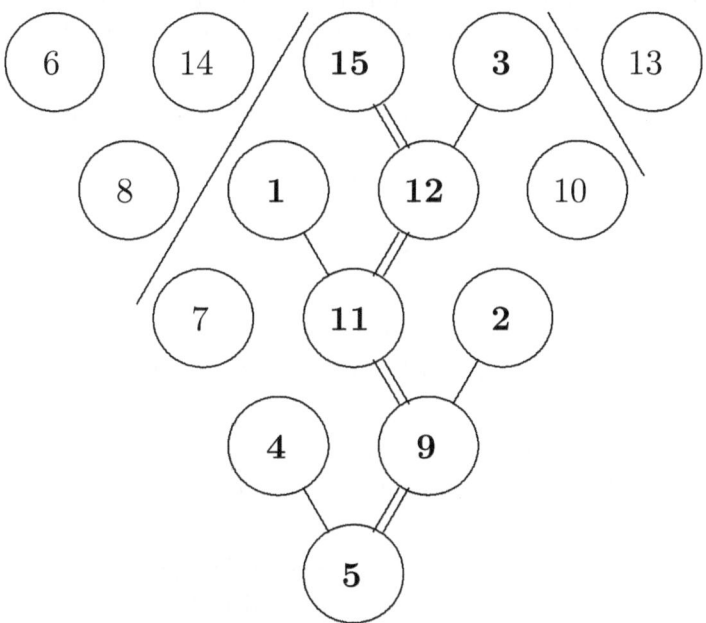

Figure 6.7

A TOAD of order n has a spine consisting of n numbers from foot to head and $n-1$ hands. The head is equal to the sum of the foot and all the hands. Since the head is at most $\frac{n(n+1)}{2}$ while the sum is at least $1+2+\cdots+n$, the head must be equal to $\frac{n(n+1)}{2}$ and the foot and all the hands must collectively be 1, 2, ..., n.

Note that there is exactly one hand or foot on each row, so that the hand or foot is the smallest number of the row. The head is clearly the largest number of the top row. It follows that the largest number on each row is on the spine.

From the head and the top hand, if we draw two slanting lines towards the sides of the TOAD, as shown in the diagram above, we can divide the TOAD into three parts. The two triangles on the side are called quasi-TOADs, with their own feet, spines, heads and hands. They are TOADs except for the fact that they do not have property 4 of the definition of a TOAD.

In the diagram above, the quasi-TOAD on the right is of order 1. Its foot and head coincide (number 13), and it has no hands. The quasi-TOAD on the left is of order 2. The spine connects the foot (number 8) directly to the head (number 14), and it has a single hand (number 6).

For $n \geq 3$, the quasi-TOADs of a TOAD of order n have spines of combined length $n - 2$, and a total of $n - 2$ feet and hands. This is still true even if one of the quasi-TOADs is empty, which happens if the spine runs along one side of the TOAD.

Suppose we have a TOAD of order $n \geq 3$. The minimum values of the $n - 2$ feet and hands of the quasi-TOADs are $n + 1$, $n + 2$, $\ldots$, $2n - 2$. Hence the sum of their heads is at least

$$(n + 1) + (n + 2) + \cdots + (2n - 2) = \frac{(n - 2)(3n - 1)}{2} = \frac{3n^2 - 7n + 2}{2}.$$

If there is only one quasi-TOAD, its head is at most $\frac{n(n+1)}{2} - 1$. From $\frac{3n^2 - 7n + 2}{2} \leq \frac{n^2 + n - 2}{2}$, we have $n^2 \leq 4n - 2$. This holds if and only if $n \leq 3$.

On the other hand, if there are two quasi-TOADs, the sum of their heads is at most $\frac{n(n+1)}{2} - 1 + \frac{n(n+1)}{2} - 2$. From $\frac{3n^2 - 7n + 2}{2} \leq n^2 + n - 3$, we have $n^2 \leq 9n - 8$. This holds if and only if $n \leq 8$. It follows that TOADs of order 9 or higher do not exist.

This proof was found in 1977 by Herbert Taylor who communicated it to Martin Gardner [6]. It has not been published until now.

For $n = 8$, the inequality $n^2 \leq 9n - 8$ obtained earlier becomes an equality. If a TOAD of order 8 exists, its head is 36 and its foot and hands are 1, 2, $\ldots$, 8. It follows that we have two quasi-TOADs. Their heads are 35 and 34, and their feet and hands are 9, 10, $\ldots$, 14. Thus each quasi-TOAD must be of order 3.

Consider the bottom three rows of the TOAD. Let the foot be x, the hand on the second row from the bottom y and the hand on the third row from the bottom z.

Without loss of generality, we may assume that the spine starts off towards the left. There are two cases, according to where the spine intersects the third row from the bottom.

Case 1. The spine continues towards the left. (See Figure 6.8.)

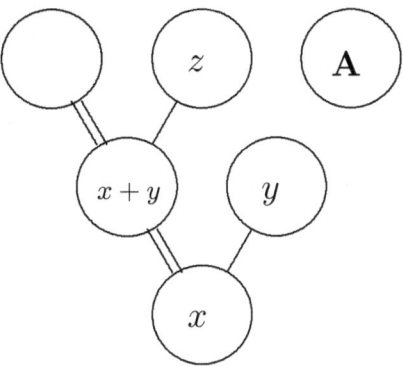

Figure 6.8

Since the numbers 1 to 14 form the hands and feet of the TOAD and the two quasi-TOADs, the number $x + y$ must be at least 15. Hence one of x and y is 8 and the other is 7, so that $z \leq 6$. Since $z < y$, we must have A$=y + z \leq 14$. This is a contradiction since A is neither a foot nor a hand of either the TOAD or one of the quasi-TOADs. Thus this case does not yield a TOAD of order 8.

Case 2. The spine turns towards the right. (See Figure 6.9.)

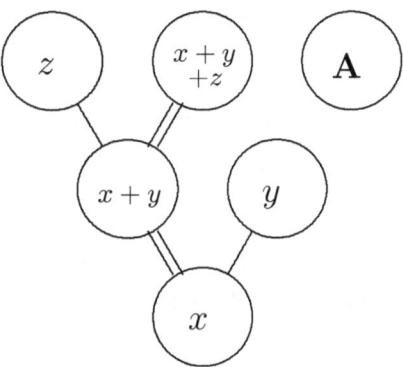

Figure 6.9

As in Case 1, one of x and y is 8 and the other is 7, so that $z \leq 6$. Since the number on the spine is the largest of the row, we must have $A=(x+y+z)-y = x+z \leq 14$. We have the same contradiction as before. Thus this case does not yield a TOAD of order 8 either.

When $n = 7$, the inequality $n^2 \leq 9n - 8$ is no longer tight. Nevertheless, if a TOAD of order 7 exists, we can still make deductions about its structure. Its head is 28 and its foot and hands are 1, 2, $\ldots$, 7. Now 8+9+10=27. It follows that one of the quasi-TOADs is of order 3, with 27 as its head and 8, 9 and 10 as its foot and hands, and the remaining two numbers chosen from 17, 18 and 19. Without loss of generality, we may assume that this quasi-TOAD is on the left side, as shown in Figure 6.10. We consider two cases.

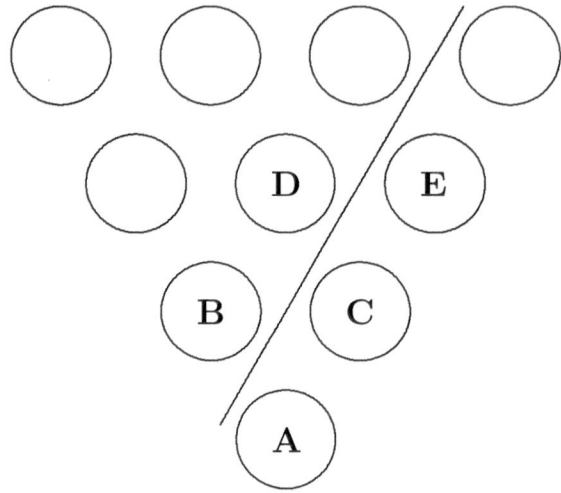

Figure 6.10

Case 1. $C \geq 19$.
Then $C \geq D \geq 8$. Hence E=27 or 28. However, these numbers had been reserved for the heads of the TOAD and the quasi-TOAD of order 3. We have a contradiction.

Case 2. $C \leq 18$. (See Figure 6.11.)
Since B=8, 9 or 10, A must be a hand of the TOAD. It follows that the spine of the TOAD must start off as shown in the diagram above, where x is the foot of the TOAD. Since $z - y < 7$, F=$y + z$. Since the head is the largest number on the top row, G=$x + y + w$ and H=$x + w - z$. Each of $y + z$, $x + w - z$ and $x + y$ is at most 13. Since 8, 9 and 10 are the foot and hands of the quasi-TOAD of order 3, these three numbers must be 11, 12 and 13 in some order. However, this means that both the foot and the hand of the quasi-TOAD of order 2 are at least 14, but its head is at most 26. We have a contradiction.

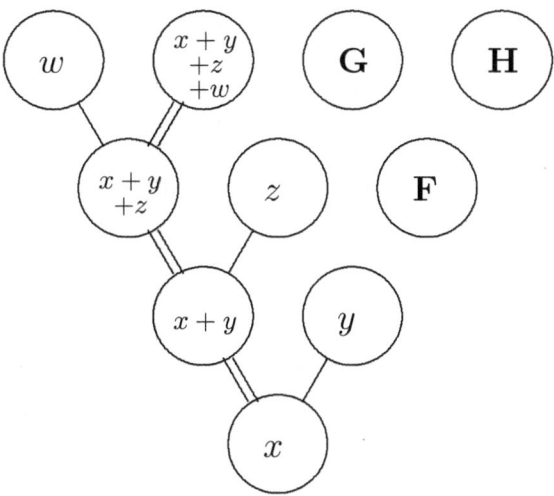

Figure 6.11

For $n = 6$, we abandon the anatomy of a TOAD and turn to the proof found in 1976 by George [2]. We work with arithmetic modulo 2 so we may replace differences by sums. The first six rows of the reduced Pascal's Triangle are shown in Figure 6.12.

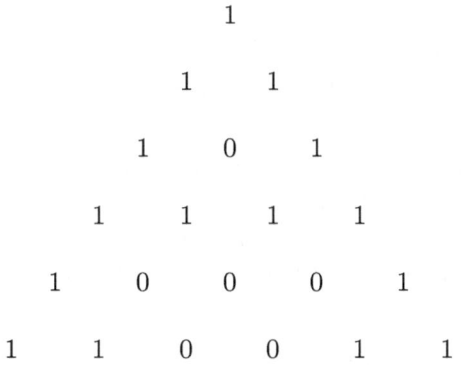

Figure 6.12

Suppose an order 6 TOAD exists. Let the numbers on the top row be a, b, c, d, e and f in that order. Then the numbers in the second row are $a + b$, $b + c$, $c + d$, $d + e$ and $e + f$. In modulo 2, those in the third row are $a + c$, $b + d$, $c + e$ and $d + f$, those in the fourth are $a + b + c + d$, $b + c + d + e$ and $c + d + e + f$, those in the fifth $a + e$ and $b + f$, and the only number in the sixth row is $a + b + e + f$. Hence their sum is $6a + 8b + 8c + 8d + 8e + 6f$, which is even. Since $1 + 2 + \cdots + 21 = 231$ is odd, we have a contradiction.

Our study concludes with the consideration of TOADs of order 5 or lower. The unique TOAD of order 1 is trivial, and there are two elementary TOADs of order 2. An order 3 TOAD has an order 1 quasi-TOAD which must be the number 4 or 5. If it is 4, this leads to the TOADs with top rows (4,1,6) and (1,6,4). If it is 5, this leads to the TOADs with top rows (5,2,6) and (2,6,5). Hence there are only four TOADs of order 3. They are shown in Figure 6.6.

We have proved earlier that any TOAD of order 4 or more must have two non-empty quasi-TOADs. Thus an order 4 TOAD has two quasi-TOADs of order 1. There are four cases.

Case 1. The top hand of the TOAD is 1.

Then 9 is not in the top row, so that 8 must be one of the quasi-TOADs. Whether it appears next to 1 or 10, 7 will not appear in the top row. Hence the other quasi-TOAD must be 6. This leads to the TOADs with top rows (6,1,10,8) and (6,10,1,8) respectively.

Case 2. The top hand is 2.

Then one of the quasi-TOAD must be 9. Whether it appears next to 2 or 10, 8 and 7 will not appear in the top row so that the other quasi-TOAD must be 6. It is routine to verify that we will need two copies of 4 to complete the TOAD.

Case 3. The top hand is 3.

Then the quasi-TOADs must be 8 and 9. This leads to the TOADs with top rows (8,3,10,9) and (8,10,3,9) respectively.

Case 4. The top hand is 4. Then all of 7, 8 and 9 need to be quasi-TOADs, which is clearly impossible.

Hence there are only four TOADs of order 4. These are shown in Figure 6.6.

An order 5 TOAD has a quasi-TOAD of order 1 and a quasi-TOAD of order 2. The latter comes from either 8+6=14 or 7+6=13. We may assume that it is on the left side, and let H be its head. There are four cases, according to the positions of H and 15. Note that 15−H=1 or 2.

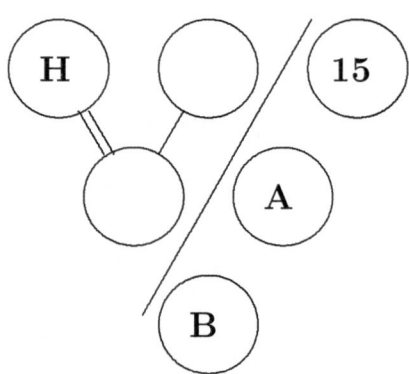

Figure 6.13

Case 1. H and 15 are in the first and third positions from the left. (See Figure 6.13.)

A must be 7, 8 or 9, and cannot be a hand. It follows that B cannot be a hand either. However, B must be 1 or 2. This is a contradiction.

Case 2. H and 15 are in the second and fourth positions from the left. (See Figure 6.14.)

C is either 1 or 2 and is therefore a hand. In order for B=A−v not to be a hand, we must have A=8 and v=1, but then H=14 and 14 will appear again in the second row from the top. Hence B and C are both hands, a contradiction.

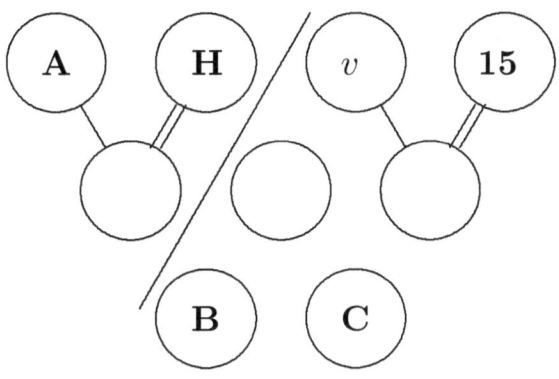

Figure 6.14

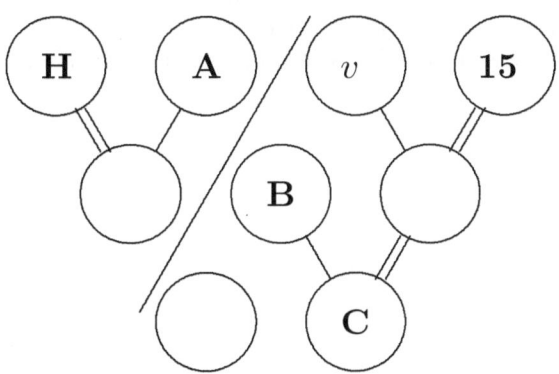

Figure 6.15

Case 3. H and 15 are in the first and fourth positions from the left. (See Figure 6.15.)

As in Case 2, B=A−v must be a hand, so that C is on the spine. Now C=15−A must be 9 as otherwise there is no possible value for the number on the spine below it. Now all of the numbers from 10 to 15 must appear in the top two rows, and we are one space short since A, v, H−A and B are all less than 9.

Case 4. H and 15 are in the second and third positions from the left. (See Figure 6.16.)

We have B=1 or 2, and as in Case 3, C=10, 11 or 12. Routine checking yields the unique solution shown in Figure 6.7.

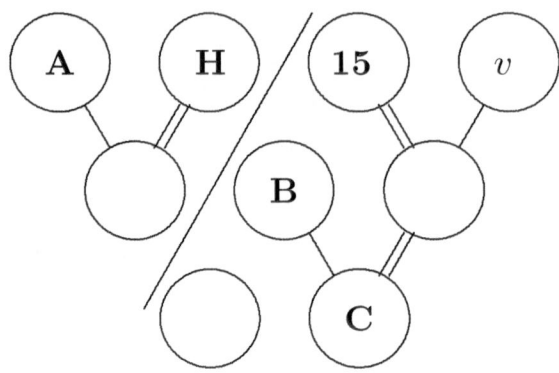

Figure 6.16

This work [3] was presented in 2010 at the Ninth Gathering for Gardner by Brian Chen, who was then in Grade 6. Two years earlier, he obtained an alternative proof for the non-existence of order 6 TOADs. Two years later, he won a Gold Medal in the International Mathematical Olympiad in Argentina.

Exercises

1. Prove directly that $\binom{k+m-1}{k}$ is the number of distinct labels of length less than m, with non-descending terms such that the first is 1 and none exceeds k.

2. Find and prove a simpler Diamond Formula for Rascal's Triangle.

3. Give an alternative proof for the non-existence of an order 6 TOAD.

Bibliography

[1] Alif Anggoro, Eddy Liu and Angus Tulloch, Rascal's Triangle, College Mathematics Journal **41** (2010) 393–395.

[2] Robert Barrington Leigh and Richard Travis Ng, Minimizing Aroma Loss, College Mathematics Journal **30** (1999) 356–358.

[3] Brian Chen, YunHao Fu, Andy Liu, George Sicherman, Herbert Taylor and Po-Sheng Wu, Triangles of Absolute Difference, submitted to *Ninth Gathering for Gardner Exchange Book*, 2010.

[4] Martin Gardner, *Penrose Tiles to Trapdoor Ciphers*, second edition, Mathematical Association of America, Washington, (1997) 119–120 and 128–129. See also Martin's *The Colossal Book of Short Puzzles and Problems*, W. W. Norton & Co., New York, (2006) 16–17 and 36–38.

[5] George Sicherman, private communication to Martin Gardner, 1976.

[6] Herbert Taylor, private communication to Martin Gardner, 1977.

Chapter Seven: Summation Problems

Section 1. Sums of Powers of Two

We consider some problems on expressing positive integers as sums of powers of 2.

Problem 1.

In how many ways can we express a positive integer n as a sum of powers of 2, if distinct powers may not be used more than once?

The number of ways is 1 for $n = 1$, namely, 1 itself. The number of ways is 1 for $n = 2$, namely, 2. The number of ways for $n = 3$ is also 1, namely, 2+1. Let the number of expressions for n be a_n. After some experimentation, we notice a pattern. We have $a_1 = 1$, $a_2 = 1$, $a_3 = 1$, $a_4 = 1$, $a_5 = 1$, $a_6 = 1$, $a_7 = 1$, and so on. We can also throw in $a_0 = 1$ for the empty sum. The pattern suggests that the general formula is $a_n = 1$. We give a formal proof via two auxiliary results.

Lemma 1. $a_{2k+1} = a_{2k}$.
Proof:
Every expression for $2k + 1$ must contain a 1. The exclusion of this 1 yields an expression for $2k$. On the other hand, every expression for $2k$ contains no 1s. The inclusion of a 1 yields an expression for $2k + 1$. This one-to-one correspondence establishes the desired result.

Lemma 2. $a_{2k} = a_k$.
Proof:
Every expression for $2k$ contains no 1s. Dividing each term by 2 yields an expression for k. This process is clearly reversible, and the one-to-one correspondence yields the desired result.

We now prove by mathematical induction that $a_n = 1$. Note that $a_0 = 1$. Suppose the result holds up to $a_{n-1} = 1$. If $n = 2k+1$, then $a_{2k+1} = a_{2k}$ by Lemma 1, and $a_{2k} = 1$ since $2k \leq n-1$. If $n = 2k$, then $a_{2k} = a_k$ by Lemma 2, and $a_k = 1$ since $k \leq n - 1$. This completes the inductive argument.

Since the pattern for a_n is relatively simple, we seek an alternative approach to Problem 1. Let $A(x) = \sum_{n=0}^{\infty} a_n x^n$. This is called the **generating function** of the sequence $\{a_n\}$. Note that x does not represent any numerical value but serves as a counter. The general term $a_n x^n$ means that the number of ways of expressing n as a sum of powers of 2 is a_n.

© Springer International Publishing AG 2018
A. Liu, *S.M.A.R.T. Circle Projects*, Springer Texts
in Education, DOI 10.1007/978-3-319-56811-9_7

What does $A(x^2) = \sum\limits_{n=0}^{\infty} a_n x^{2n}$ generate? The general term $a_n x^{2n}$ means
that the number of ways of expressing $2n$ as a sum of powers of 2 with some
restrictions is a_n, which is the number of ways of expressing n as a sum of
powers of 2. From Lemma 2, we see that this restriction is that we are not
allowed to use 1. Similarly $x A(x^2) = \sum\limits_{n=0}^{\infty} a_n x^{2n+1}$ generates the number of
ways of expressing $2n + 1$ with exactly one 1. It follows that

$$
\begin{aligned}
A(x) &= (1+x)A(x^2) \\
&= \frac{1-x^2}{1-x} A(x^2) \\
&= \frac{1-x^2}{1-x} \cdot \frac{1-x^4}{1-x^2} A(x^4) \\
&= \cdots \\
&= \frac{1}{1-x}.
\end{aligned}
$$

Now $\frac{1}{1-x} = 1 + x + x^2 + x^3 + \cdots$. Hence $a_n = 1$ for all n.

Problem 2.

In how many ways can we express a positive integer n as a sum of powers
of 2, if distinct powers may not be used more than twice?

Let the number of expressions for n be $b_n = b(n)$. The first 16 values are

$$
\begin{array}{llll}
b_1 = 1 & b_2 = 2 & b_3 = 1 & b_4 = 3 \\
b_5 = 2 & b_6 = 3 & b_7 = 1 & b_8 = 4 \\
b_9 = 3 & b_{10} = 4 & b_{11} = 2 & b_{12} = 4 \\
b_{13} = 3 & b_{14} = 4 & b_{15} = 1 & b_{16} = 5
\end{array}
$$

We can also throw in $b_0 = 1$ for the empty sum.

The general pattern is unclear, but we make some observations. For
$t \geq 1$, we have

$$
\begin{aligned}
b(2^t - 1) &= 1, & (6) \\
b(2^t) &= t + 1, & (7) \\
b(2^t + 1) &= t, & (8) \\
b(3(2^{t-1}) - 1) &= 2. & (9)
\end{aligned}
$$

For a general approach, we first prove two auxiliary results.

Lemma 3. $b_{2k+1} = b_k$.

Proof:

The proof is exactly the same as that for Lemma 1.

Lemma 4. $b_{2k} = b_k + b_{k-1}$.
Proof:
Every expression for $2k$ contains either no 1s or two 1s. In the former case, dividing each term by 2 yields an expression for k. In the latter case, dividing each term by 2 after the exclusion of the two 1s yields an expression for $n - 1$. These two processes are clearly reversible, and the one-to-one correspondences yield the desired result.

A general formula for b_n does not seem to be within reach. The method of generating functions does not work here either. However, with Lemma 3 and Lemma 4, we can compute the value of b_n for any positive integer n. We can also prove that the observations are indeed correct.

We shall establish (1) and (2) by simultaneous induction. For $t = 1$, we have $b(2 - 1) = 1$ and $b(2) = 2$. Suppose (1) and (2) hold for some $t \geq 1$. Then $b(2^{t+1} - 1) = b(2(2^t - 1) + 1) = b(2^t - 1) = 1$ by Lemma 3, and $b(2^{t+1}) = b(2(2^t)) = b(2^t) + b(2^t - 1) = t + 1$ by Lemma 4.

We can also establish (3) and (4) by mathematical induction. For $t = 1$, we have $b(2 + 1) = 1$ and $b(3(1) - 1) = 2$. Suppose (3) and (4) hold for some $t \geq 1$. By Lemma 3, we have

$$b(2^{t+1} + 1) = b(2(2^t) + 1) = b(2^t) = t + 1$$

and

$$b(3(2^t) - 1) = b(2(3(2^{t-1} - 1) + 1) = b(3(2^{t-1} - 1) = 2.$$

Note also the palindrome blocks (1,2,1), (1,3,2,3,1), (1,4,3,4,2,4,3,4,1), ... For $t \geq 1$, they are centered at $b(3(2^{t-1}) - 1)$ and extend from $b(2^t - 1)$ to $b(2^{t+1} - 1)$.

Problem 3.
In how many ways can we express a positive integer n as a sum of powers of 2, if distinct powers may not be used more than thrice?

Let the number of expressions for n be c_n. The first 7 values are 1, 2, 2, 3, 3, 4 and 4. We can also throw in $c_0 = 1$ for the empty sum. The pattern suggests that the general formula is $c_{2n} = c_{2n+1} = n + 1$. A slightly more compact way to express this result is $c_n = \lfloor \frac{n+2}{2} \rfloor$.

As before, we first prove two auxiliary results.

Lemma 5. $c_{2k+1} = c_{2k}$.
Proof:
Every expression for $2k + 1$ must contain at least one 1. The exclusion of one 1 yields an expression for $2k$. On the other hand, every expression for $2k$ can contain at most two 1s. The inclusion of one 1 yields an expression for $2k + 1$. This one-to-one correspondence establishes the desired result.

Lemma 6. $c_{2k} = c_k + c_{k-1}$.

Proof:

The proof is exactly the same as that for Lemma 4.

We now prove by mathematical induction that $c_{2n} = c_{2n+1} = n + 1$. Note that $c_0 = c_1 = 1$. Suppose the result holds up to $c_{2n-2} = c_{2n-1} = n$. By the Lemmas, we have $c_{2n} = c_{2n+1} = c_n + c_{n-1}$. Suppose $n = 2k$. Then $c_n + c_{n-1} = c_{2k} + c_{2k-1} = k + 1 + k = 2k + 1 = n + 1$. Suppose $n = 2k + 1$. Then $c_n + c_{n-1} = c_{2k+1} + c_{2k} = k + 1 + k + 1 = 2k + 2 = n + 1$. This completes the inductive argument.

Using the generating function approach, let $C(x) = \sum_{n=0}^{\infty} c_n x^n$. Then $C(x^2)$ generates the number of expressions with no 1s, $xC(x^2)$ generates the number of those with one 1, $x^2 C(x^2)$ generates the number of those with two 1s and $x^3 f(x^2)$ generates the number of those with three 1s. It follows that

$$
\begin{aligned}
C(x) &= (1 + x + x^2 + x^3)C(x^2) \\
&= \frac{1 - x^4}{1 - x} C(x^2) \\
&= \frac{1 - x^4}{1 - x} \cdot \frac{1 - x^8}{1 - x^2} C(x^4) \\
&= \frac{1 - x^4}{1 - x} \cdot \frac{1 - x^8}{1 - x^2} \cdot \frac{1 - x^{16}}{1 - x^4} C(x^8) \\
&= \cdots \\
&= \frac{1}{(1 - x)(1 - x^2)} \\
&= \frac{\alpha}{1 - x} + \frac{\beta}{(1 - x)^2} + \frac{\gamma}{1 + x}.
\end{aligned}
$$

Clearing fractions, we have $1 = \alpha(1 - x)(1 + x) + \beta(1 + x) + \gamma(1 - x)^2$. Setting $x = -1$, $1 = 4\gamma$ so that $\gamma = \frac{1}{4}$. Setting $x = 1$, $1 = 2\beta$ so that $\beta = \frac{1}{2}$. Setting $x = 0$, $1 = \alpha + \beta + \gamma$ so that $\alpha = \frac{1}{4}$. Now $\frac{1}{1-x} = 1 + x + x^2 + x^3 + \cdots$ and $\frac{1}{1+x} = 1 - x + x^2 - x^3 + \cdots$. On the other hand,

$$
\begin{aligned}
\frac{1}{(1 - x)^2} &= (1 + x + x^2 + x^3 + \cdots)^2 \\
&= 1 + 2x + 3x^2 + 4x^3 + \cdots.
\end{aligned}
$$

Hence $c_n = \frac{1}{2}(n + 1) + \frac{1}{4}(1 + (-1)^n) = \lfloor \frac{n+2}{2} \rfloor$.

This work, by Circle members Dennis Situ and Steven Xia, was published as [7].

Section 2. Zig-Zag

The following problem, which has become a classic, first appeared in a Hungarian competition [5].

Determine which positive integers cannot be expressed as sums of two or more consecutive positive integers.

Testing the numbers from 1 to 10, we have 3=1+2, 5=2+3, 6=1+2+3, 7=3+4, 9=4+5=2+3+4 and 10=1+2+3+4. The missing numbers are 1, 2, 4 and 8. It is reasonable to conjecture that the powers of 2 are the *undoable* numbers we were looking for.

All odd numbers $2n + 1$ where $n > 0$ are *doable* because it is equal to $n + (n + 1)$. All numbers of the form $\frac{n(n+1)}{2}$ where $n > 1$ are also doable because it is equal to $1 + 2 + \cdots + n$. Neither of the two expressions for the number 9 starts with 1. Still, we can draw a pair of staircases which will form a box, as shown in Figure 7.1. By a *box*, we mean a rectangle with integer dimensions *greater* than 1.

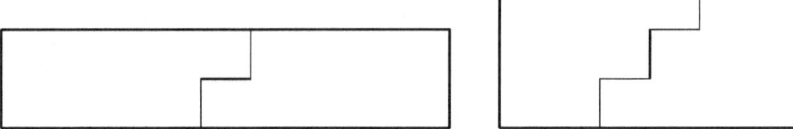

Figure 7.1.

Each box is divided into halves by a *central zigzag*. Now any such zigzag starts with a vertical *zag* followed by a horizontal *zig*. Then comes another zag followed by a zig, and so on. It ends with a zag, but it is not followed by a zig. This means that the total number of zigzags is odd. It follows that the central zigzag has a middle zig or zag. We have a middle zig in one diagram for the number 9, and a middle zag in the other. This does not really matter.

If we have a middle zig, then the horizontal dimension of the box must be odd. This is because the two staircases are symmetric about this middle zig. Similarly, if we have a middle zag, then the vertical dimension of the box must be odd. Thus a positive integer n is doable if we can find a box of area $2n$ with one of its dimensions odd.

Let us say that it is the horizontal one. Then we have a middle column. Because the area of the box is even, the vertical dimension must be even. Then we have a horizontal line which divides the box into halves. Where this line meets the middle column is the middle zig for our central zigzag. It is now easy to construct the central zigzag by extending the middle zig in opposite directions. The situation is analogous if the vertical dimension of the box is odd.

If there is such a box, n must have an odd divisor greater than 1. This is the case if n is not a power of 2. Thus all numbers which are not powers of 2 can be expressed as the sum of two or more consecutive positive integers. Moreover, the number of such expressions for n is the same as the number of odd factors of n which are greater than 1. This is because different boxes give us different expressions.

What happens if n is a power of 2? Then $2n$ is also a power of 2. In any box with area $2n$, both dimensions are even. So there will not be a middle column or a middle row. Thus there is nothing that can serve as the middle zig or zag of a central zigzag, and it is impossible to divide the box into two identical staircases. Therefore, a power of 2 cannot be expressed as a sum of two or more consecutive integers.

This beautiful geometric solution was the work of Circle members Robert Barrington Leigh and Richard Ng, who were in grades 6 and 7 respectively at the time. Their work was published as [1].

It is possible to give a completely algebraic treatment to this problem. Suppose we have an expression $(a+1) + (a+2) + \cdots + b$. We can rewrite it as the difference between $1 + 2 + \cdots + b$ and $1 + 2 + \cdots + a$. Therefore, the value of our expression is $\frac{b(b+1)}{2} - \frac{a(a+1)}{2}$. It can be factored into $\frac{(b-a)(b+a+1)}{2}$. Since $b + a$ and $b + a + 1$ are consecutive, one is odd and the other is even. Since $(b+a) + (b-a) = 2b$, either both are even or both are odd. It follows that one of $b - a$ and $b + a + 1$ is odd. Their product is $2n$, and it can only be a power of 2 if the odd factor is 1. Obviously, $a + b + 1 > 1$. If $b - a = 1$, then $a + 1 = b$, and $(a+1) + (a+2) + \cdots + b$ reduces to a single term b. It follows that a power of 2 is undoable.

Suppose $n = cd$ where $c > 1$ is odd. We can write down c copies of the number d. Since c is odd, there is a middle number. Keep it as d. On one side, replace the d's by $d + 1, d + 2, \ldots, d + \frac{c-1}{2}$, and on the other side, replace the d's with $d - 1, d - 2, \ldots, d - \frac{c-1}{2}$. If some of these numbers become negative, there will be a 0 and positive numbers immediately on the other side of 0 which will cancel the negative numbers off, so that the string of consecutive numbers is not broken. For example, we have

$$9 = 3 + 3 + 3 = 2 + 3 + 4$$

and

$$9 = 1+1+1+1+1+1+1+1+1 = (-3)+(-2)+(-1)+0+1+2+3+4+5 = 4+5.$$

Since $c > 1$ and $d > \frac{1}{2}$, we have $d(c-1) > \frac{c-1}{2}$ and $cd > d + \frac{c-1}{2}$. Hence the cancellation cannot reduce the expression to a single term.

Section 3. Two Great Escapes.

Having encountered arithmetic progressions in the previous sections, it is natural to ask whether there is any sequence called a **geometric progression**. Indeed there is. In fact, we have met a prototype already, namely, the powers of 2. Here, the first term is arbitrary. Each subsequent term is obtained from the preceding one by multiplying with a constant called the **common ratio**. The special case where the common ratio is 1 yields a constant sequence. The special case where the common ratio is 0 yields a sequence in which every term except possibly the first one is 0.

Here are some examples of geometric progressions.

(1) 1, 3, 9, 27, 81, 243, ...
(2) 1, −1, 1, −1, 1, −1 ....
(3) 4, 2, 1, $\frac{1}{2}$, $\frac{1}{4}$, $\frac{1}{8}$, ...
(4) −6, 3, −1$\frac{1}{2}$, $\frac{3}{4}$, −$\frac{3}{8}$, $\frac{3}{16}$, ...

The summation method will be introduced in the following two adventures.

The Great Amoeba Escape

The world of the amoeba consists of the first quadrant of the plane divided into unit squares. Initially, a solitary amoeba is imprisoned in the bottom left corner square. The prison consists of the six shaded squares as shown in Figure 7.2. It is unguarded, and the Great Escape is successful if the entire prison is unoccupied.

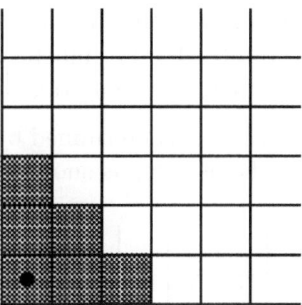

Figure 7.2.

In each move, an amoeba splits into two, with one going to the square directly north and one going to the square directly east. However, the move is not permitted if either of those two squares is already occupied. Can the Great Escape be achieved?

After a few moves, we arrive at the situation in Figure 7.3, where only one amoeba is still in jail. However, we are already in a very heavy traffic jam, and it is not clear whether this last one can get out. So we need to work out some sort of strategy.

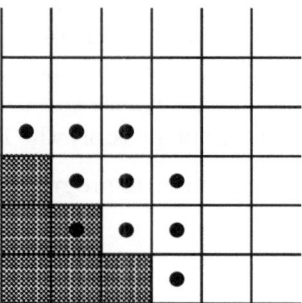

Figure 7.3.

Note that the configuration keeps changing, with more and more amoebas. The changes must be carefully monitored before things get out of hand. What we seek is a quantity which remains unchanged throughout. Such a quantity is called an *invariant*.

At the start, we have only one amoeba. After one move, we have two amoebas. However, each is really less than one full amoeba. Suppose we assign the value 1 to the initial amoeba, x to the one going north and y to the one going east. After the move, the initial amoeba is replaced by the other two. If we want the total value of amoebas to remain constant, we must have $x + y = 1$. By symmetry, we may take $y = x$ so that $x = \frac{1}{2}$.

Clearly, the value of an amoeba is determined by its location. So we may assign values to the squares themselves, as shown in Figure 7.4.

$\frac{1}{16}$	$\frac{1}{32}$	$\frac{1}{64}$	$\frac{1}{128}$	$\frac{1}{256}$
$\frac{1}{8}$	$\frac{1}{16}$	$\frac{1}{32}$	$\frac{1}{64}$	$\frac{1}{128}$
	$\frac{1}{8}$	$\frac{1}{16}$	$\frac{1}{32}$	$\frac{1}{64}$
		$\frac{1}{8}$	$\frac{1}{16}$	$\frac{1}{32}$
			$\frac{1}{8}$	$\frac{1}{16}$

Figure 7.4.

The total value of the squares in the first row is

$$S = 1 + \frac{1}{2} + \frac{1}{4} + \frac{1}{8} + \cdots.$$

This is a geometric progression. We have

$$2S = 2 + 1 + \frac{1}{2} + \frac{1}{4} + \frac{1}{8} + \cdots.$$

Subtracting the previous equation from this one, we have $S = 2$. Since each square in the second row is half in value of the corresponding square in the first row, the total value of the squares in the second row is 1. Similarly, the total values of the squares in the remaining rows are $\frac{1}{2}$, $\frac{1}{4}$, $\frac{1}{8}$, ... Hence the total value of the squares in the entire quadrant is 4.

Note that the total value of the six prison squares is $2\frac{3}{4}$. Remember that the total value of the amoebas is the invariant 1. If the Great Escape is to be successful, the amoebas must fit into the non-prison squares with total value $1\frac{1}{4}$. While there is no immediate contradiction, we do not have much room to play about.

The amount of amoeba that the first row can hold outside of the prison is $\frac{1}{8} + \frac{1}{16} + \frac{1}{32} + \cdots = \frac{1}{4}$, and the same amount can be held in the first column outside of the prison. Thus the remaining space can hold only $\frac{3}{4}$ of an amoeba, as shown in Figure 7.5.

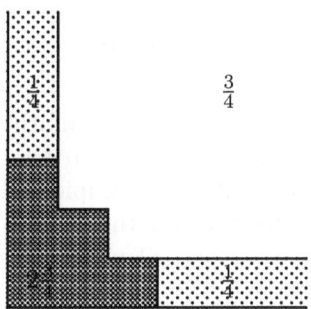

Figure 7.5.

Each of the first row and the first column holds exactly one amoeba at any time. If the amoeba on the first row is outside the prison, its value is at most $\frac{1}{8}$. The remaining space with total value $\frac{1}{4} - \frac{1}{8} = \frac{1}{8}$ must be wasted. Similarly, we have to leave vacant squares in the first column with total value at least $\frac{1}{8}$. Since $1\frac{1}{4} - 2 \times \frac{1}{8} = 1$, we have no room to play at all.

In order for the Great Escape to be successful, all squares outside the prison and not on the first row or first column must be occupied. However, this requires that the number of moves must be infinite. Hence the Great Escape cannot be achieved in a finite number of moves.

The Great Beetle Escape

The world of the beetles consists of the entire plane divided into unit squares. Initially, all squares south of a wall constitute the prison, and each is occupied by a beetle. There is a corridor of length 5 extending beyond the wall. On the last square, which is shaded in Figure 7.6, there is a special telephone. If any beetle reaches there, it can call the Great Beetle in the Sky, who will then come to release all beetles still in prison. In that case, the Great Escape is successful.

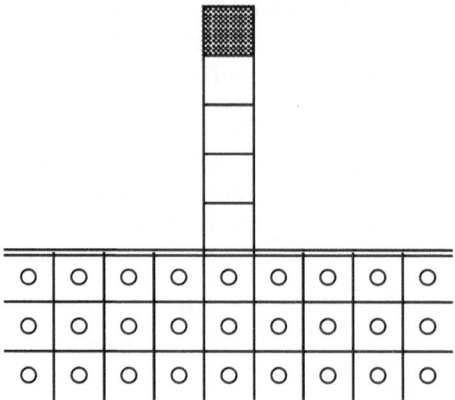

Figure 7.6.

In each move, a beetle can jump over another beetle in an adjacent square and land on the square immediately beyond. However, the move is not permitted if that square is already occupied. The beetle being jumped over is removed, making a sacrifice for the common good. The jump may be northward, eastward or westward. Can the Great Escape be achieved?

Again, we seek an invariant. Assign 1 to the beetle at the telephone at the end. It gets to its present position by jumping over another beetle. Assign x to that beetle and y to the beetle before making the jump. After the move, the final beetle replaces the other two. In order for the total value of the beetles to be invariant, we must have $x + y = 1$.

Now a beetle with value z could jump over the one of value y to become the one with value x. If we choose $y = x$ as in the Amoeba Problem, then we must take $z = 0$ in order to maintain $z + y = x$. This is undesirable. A better choice is $y = x^2$. Then we can take $z = x^3$. Since $x^2 + x = 1$, we indeed have $z + y = x^3 + x^2 = x(x^2 + x) = x$. By the Quadratic Formula, the positive root of $x^2 + x - 1 = 0$ is given by $x = \frac{\sqrt{5}-1}{2} \approx 0.618$. It is known as the Golden Ratio.

The value of a beetle is also determined by its location. So we may assign values to the squares themselves, as shown in Figure 7.7.

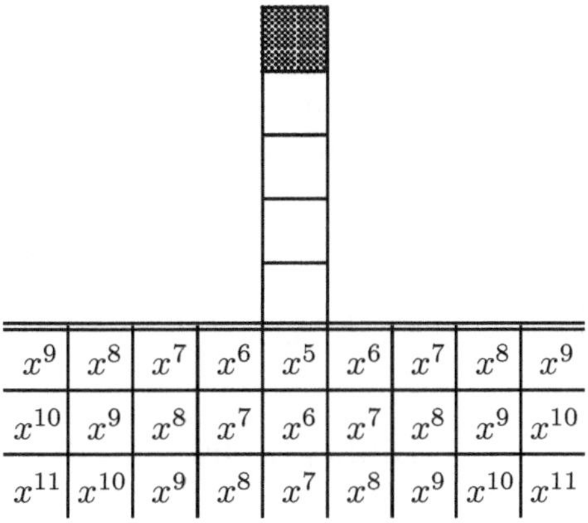

x^9	x^8	x^7	x^6	x^5	x^6	x^7	x^8	x^9
x^{10}	x^9	x^8	x^7	x^6	x^7	x^8	x^9	x^{10}
x^{11}	x^{10}	x^9	x^8	x^7	x^8	x^9	x^{10}	x^{11}

Figure 7.7.

The total value of the squares in the prison is

$$S = x^5 + 3x^6 + 5x^7 + 7x^8 + \cdots.$$

We have

$$xS = x^6 + 3x^7 + 5x^8 + 7x^9 + \cdots.$$

Subtracting this equation from the previous one, we have

$$
\begin{aligned}
(1 - x)S &= x^5 + 2(x^6 + x^7 + x^8 + \cdots) \\
&= x^5 + \frac{2x^6}{1 - x} \\
&= \frac{x^5 + x^6}{1 - x}.
\end{aligned}
$$

It follows that $S = \frac{x^5 + x^6}{(1-x)^2}$. Recall that $x^2 + x = 1$, so that $1 - x = x^2$. Hence the denominator of S is $(1 - x)^2 = (x^2)^2 = x^4$. The numerator of S is $x^6 + x^5 = x^4(x^2 + x) = x^4$ also, so that $S = 1$. Thus the Great Escape can only be successful by sacrificing all but one beetle, and cannot be achieved in a finite number of moves.

This section is based on the paper [6] by Circle member J. Lo. The Great Amoeba Escape is due to M. Kontsevich (see [9]) and the Great Beetle Escape is due to J. H. Conway (see [4]). The idea of an invariant is an important problem-solving technique. For further discussions and practices, see [2], [3] and [8].

Exercises

1. How many of the first 100 positive integers can be expressed as a sum of powers of 2 if no two identical or consecutive powers of 2 are used?

2. The number 336 has an odd divisor 7, and 7=3+4. Use this as a starting point to find an expression of 336 as a sum of consecutive positive integers.

3. How many beetles do we need to accomplish the Great Beetle Escape if the length of the corridor is
   (a) 1;   (b) 2;   (c) 3;   (d) 4?

Bibliography

[1] Robert Barrington Leigh and Richard Ng, Zigzag, Mathematics Competitions, **10** (1997) 38-42, republished in Hungarian as Cikcakk, Abacus **4** (1998) 318–320'.

[2] A. Engel, Problem-Solving Strategies, Springer, New York, (1997) 1–23.

[3] D. Fomin, S. Genkin and I. Itenberg, Mathematical Circles, Amer. Math. Soc., Providence, (1996) 123–133, 254–257.

[4] R. Honsburger, Mathematical Gems II, Math. Assoc, Amer., Washington, (1976) 23–28.

[5] Andy Liu, "Hungarian Problem Book IV (1947–1963)", Mathematical Association of America (2001) 2 and 34–35.

[6] Jerry Lo, Two Great Escapes, Delta K, **43**-2 (2005) 23–27.

[7] Dennis Situ and Steven Xia, The Powers of Two, Mathematics Competitions, **28** (2015) #2, 37-41.

[8] Jordan Tabov and Peter Taylor, Methods of Problem Solving I, Austral. Math. Trust, Canberra, (1996) 93–109.

[9] Peter Taylor, Tournament of the Towns 1980–1984, Austral. Math. Trust, Canberra, (1993) 31, 37–39.

Chapter Eight: Finite Projective Geometries

Section 1. Lions and Ponies

Many students have difficulties with proofs. This is understandable since the concept of proofs lies in the heart of mathematics, and proofs are not the most intuitive things to do. On top of that, proofs cannot exist in a vacuum, and often, the subject matter adds to the difficulty.

What we will do now is to consider a situation which does not apparently involve any mathematical concepts. We will then do some proofs without being encumbered by having to deal with contents. Thus we can focus on the proof process itself.

A certain community of lions and ponies is defined by the following postulates.
(1) There are at least two lions.
(2) Each lion has bitten at least three ponies.
(3) For any pair of lions, there is exactly one pony that both have bitten.
(4) For any pair of ponies, there is at least one lion that has bitten both.

So far, there is nothing that would scare students, apart from being bitten by lions. Nevertheless, from this humble beginning, we can derive many interesting results.

Theorem 1.
For any lion, there is at least one pony that it has not bitten.

Theorem 1 is essentially a negative statement. To prove it, we let L be a given lion. We have to find a pony which it has not bitten. Where do we begin? We know nothing about this community except for the four postulates which define it. All of them are positive statements.

In the absence of any good ideas, we may use an indirect approach. Assume the opposite of what we have to prove, and see what may be wrong with it. So we suppose that L has bitten every pony. By Postulate 1, there is another lion M. By Postulate 3, there is exactly one pony which both L and M have bitten. Since L has bitten every pony, this means that M can have bitten only one pony.

However, Postulate 2 states clearly that M has bitten at least three ponies. What is going on? What we have is a contradiction. Where does it arise? Did we make any mistakes in our logical reasoning? Going over the steps carefully, they are all correct. Hence the problem arises from our initial assumption, that L has bitten every pony. It follows that L has not bitten every pony, and Theorem 1 has been proved.

© Springer International Publishing AG 2018
A. Liu, *S.M.A.R.T. Circle Projects*, Springer Texts
in Education, DOI 10.1007/978-3-319-56811-9_8

We can now turn this argument around and give a direct proof of Theorem 1. For a given lion L, we wish to prove that there is a pony which it has not bitten. By Postulate 1, there is another lion M. By Postulate 2, M has bitten at least three ponies P, Q and R. By Postulate 3, there is exactly one pony, say P, which both L and M have bitten. Then Q is a pony which L has not bitten.

Theorem 2.
For any pony, there is at least one lion that has not bitten it.

Theorem 2 is obtained from Theorem 1 by interchanging the roles of the lions and ponies. It is called the *dual* of Theorem 1. Of course, Theorem 1 is also the dual of Theorem 2.

Again, we start with an indirect approach. Assume that there is a pony P that every lion has bitten. By Postulate 1, there are two lions L and M. By Postulate 2, L has bitten a pony Q other than P and M has bitten a pony R other than P. By Postulate 3, L has not bitten R and M has not bitten Q. Hence Q and R are not the same pony. By Postulate 4, there is a lion N that has bitten both Q and R. Hence N and L are two different lions but they have both bitten P and Q. This contradicts Postulate 3.

Very minor changes yield a direct proof. By Postulate 1, there are two lions L and M. If one of them has not bitten P, there is nothing else to prove. So assume that they both have. By Postulate 2, L has bitten a pony Q other than P and M has bitten a pony R other than P. By Postulate 3, L has not bitten R and M has not bitten Q. Hence Q and R are not the same pony. By Postulate 4, there is a lion N that has bitten both Q and R. Hence N and L are two different lions. Since they have both bitten Q, N is a lion which has not bitten P by Postulate 3.

Theorem 3.
For any pair of lions, there is at least one pony that neither has bitten.

Let L and M be any pair of lions. By Postulate 3, there is a pony P that both L and M have bitten. By Theorem 2, there is a lion N that has not bitten P. By Postulate 2, N has bitten at least three ponies. By Postulate 3, L has bitten at most one of these three and so has M. Hence there is a pony that neither L nor M has bitten.

Theorem 4.
For any pair of ponies, there is at least one lion that has not bitten either.

Let P and Q be any pair of ponies. By Postulate 4, there is a lion L that has bitten both of them. By Postulate 2, this lion has bitten a third pony R. By Theorem 1, there is a pony S that L has not bitten. By Postulate 4, there is a lion M that has bitten both R and S. By Postulate 3, M has not bitten either P or Q.

Just as Theorems 1 and 2 are the duals of each other, Theorems 3 and 4 are the duals of each other.

We have done some reasonably sophisticated proofs without delving into any formal subject matter. Clearly, there cannot be a community in the real world where lions bite ponies according to a set of postulates. Nevertheless, one should raise the question whether such an abstract structure can in fact exist. If not, all our proofs so far are for nothing.

Before we pursue this angle, let us recall that duality is a central concept in our community of lions and ponies. Duality allows us to obtain two results for the price of one, which is a very good thing. It was this search for duality that led to a very important development in the history of mathematics.

In Euclidean geometry, the first postulate is that every two points determine a line. The dual of this result is that every two lines determine a point. It is almost true, except in the case where the lines are parallel to each other. If we wish this dual to hold true, then parallel lines must also meet at some point.

In everyday life, the two rails of a straight railway track must be parallel, as otherwise any train running on them must derail. However, in a painting or photograph of a railway track, the two rails may perhaps not come to a point, but they are most certainly not parallel. They come closer and closer to each other as they recede into the background, which leads to the saying that *parallel lines meet at infinity.*

So we add a point at infinity to each line in the Euclidean plane. We call them *dieal* points, to distinguish them from the ordinary points. Parallel lines have the same ideal point while non-parallel lines have different ideal points. In a sense, an ideal point represents the common slope of a set of parallel lines. In this extended plane, every two lines determine a point.

We have plugged one hole, but we may have cracked another leak. Is it still true that every two points determine a line?

If both points are ordinary points, then they determine a line as usual. If one point is ordinary and the other is ideal, they still determine a line. This is the familiar point-slope formula in analytic geometry. However, if both points are ideal, they do not determine a line.

To plug this new hole, we now add an *ideal* line which passes through all the ideal points and only the ideal points. So two ideal points will determine the ideal line. Moreover, the ideal line and an ordinary line determines the ideal point on that ordinary line, so we have indeed achieved duality.

This new plane is called the *projective* plane, and a new branch of mathematics called *projective* geometry is born. The quest for duality, though apparently an exercise in abstraction, does have important consequences.

The church paintings (only the churches could afford paintings in those days) up until about the twelfth century did not have depth perception. This was because the concept of perspective views had not yet existed. Paintings since then have benefited from the development of projective geometry.

We now see that if we replace the ponies by points and the lions by lines, our community becomes a projective geometry. However, it is clear that we intend only to have a finite number of lions and a finite number of ponies. So we turn our attention to the subject of *finite* geometry.

Figure 8.1 shows the simplest finite model of the Euclidean plane. There are four points A, B, C and D, with coordinates $(x, y) = (0, 0)$, $(1,0)$, $(0,1)$ and $(1,1)$ respectively There are six lines AB $(y = 0)$, CD $(y = 1)$, AC $(x = 0)$, BD $(x = 1)$, AD $(x + y = 0)$ and BC $(x + y = 1)$. Note that addition is in modulo 2, so that the point $(1,1)$ lies on $x + y = 0$.

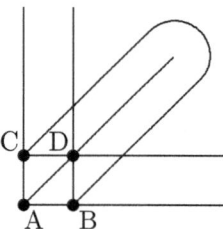

Figure 8.1

Clearly, AB and CD are parallel to each other, as are AC and BD. Moreover, AD and BC are also parallel to each other, as they do not intersect at any of the points in this geometry. We have deliberately drawn BC in such a way to emphasize this point. We call this finite plane the *affine* plane of *order* 2.

We now add an ideal point E to AB and CD, an ideal point F to AC and BD, an ideal point G to AD and BC, and an ideal line passing through E, F and G. This extended plane, shown on the left of Figure 8.2, is called the *projective* plane of order 2. It is redrawn in a more stylish form on the right of Figure 8.2, where the line BC is still drawn as a curve. This is a famous example called the *Fano* plane.

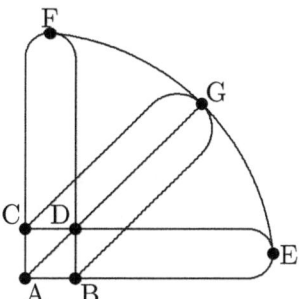 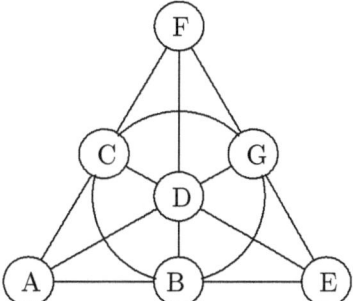

Figure 8.2

Now let the seven points represent the ponies and let the seven lines represent the lions. Each lion bites the three ponies represented by points on the line representing the lion. We can verify that the four postulates of the community of lions and ponies indeed hold. They now become the postulates of the projective geometry, and we restate them as follows.

(1) There are at least two lines.

(2) Each line passes through at least three points.

(3) For any pair of lines, there is exactly one point that both pass through.

(4) For any pair of points, there is at least one line that passes through both.

Figure 8.3 shows the projective plane of order 3, with the affine plane of order 3 embedded in it.

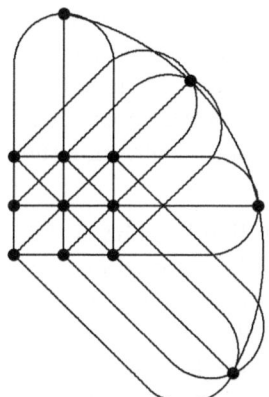

Figure 8.3

We must exercise caution in trying to construct an affine plane of order 4. It would appear natural to take the points with coordinates $(x, y) = (0, 0)$, $(0,1)$, $(0,2)$, $(0,3)$, $(1,0)$, $(1,1)$, $(1,2)$, $(1,3)$, $(2,0)$, $(2,1)$, $(2,2)$, $(2,3)$, $(3,0)$, $(3,1)$, $(3,2)$ and $(3,3)$. The lines may be divided into the following five groups of parallel lines.

$$
\begin{array}{lllll}
y = 0 & x = 0 & y = x & y = 2x & y = 3x \\
y = 1 & x = 1 & y = x+1 & y = 2x+1 & y = 3x+1 \\
y = 2 & x = 2 & y = x+2 & y = 2x+2 & y = 3x+2 \\
y = 3 & x = 3 & y = x+3 & y = 2x+3 & y = 3x+3
\end{array}
$$

Consider the line $y = 2x$ which passes through the four points $(0,0)$, $(1,2)$, $(2,0)$ and $(3,2)$. It will intersect the line $y = 0$ in two points, namely, $(0,0)$ and $(2,0)$, and this violates the first postulate of Euclidean geometry. The reason for this is that 4 is not a prime number, so that we can have $2 \times 2 \equiv 0 \pmod 4$. Nevertheless, there is an affine plane of order 4, but it is based on a concept called *Galois* theory which we will not discuss here.

In general, if n is a prime number, then we can construct an affine plane of order n in the usual way. It has n^2 points and $n^2 + n$ lines which may be divided into $n + 1$ classes of n parallel lines. Each line passes through n points and each point lies on $n + 1$ lines. By adding an ideal point to each line and joining them by an ideal line, we have a projective plane of order n. It has $n^2 + n + 1$ points and $n^2 + n + 1$ lines. Each line passes through $n + 1$ points and each point lies on $n + 1$ lines. Observe the duality between the points and the lines here.

To conclude this section, let us return to the community of lions and ponies and work on some counting problems. Let L be a lion and P be a pony such that L has not bitten P. We claim that the number of ponies that L has bitten is equal to the number of lions that have bitten P.

Let M be another lion which has bitten P and Q be another pony L has bitten. By Postulate 3 and its dual (proof left as an exercise), there is exactly one pony that both L and M have bitten and exactly one lion that has bitten both P and Q. Hence the ponies that L has bitten and the lions that have bitten P can be paired off, and their numbers are equal. This justifies our claim.

We now prove that each lion has bitten the same number of ponies. Let L and M be any two lions. We have proved earlier that there is a pony P that neither has bitten. Hence the number of ponies that L has bitten is equal to the number of ponies that M has bitten. This follows from the claim above since both are equal to the number of lions that have bitten P. In an analogous manner, we can prove that each pony has been bitten by the same number of lions.

Suppose one of the lions, say L, has bitten exactly $n + 1$ ponies. Then every lion has bitten exactly $n + 1$ ponies. Now there is a pony P that L has not bitten. It follows that exactly $n + 1$ lions has bitten P, so that each pony has been bitten by exactly $n + 1$ lions.

Consider the $n+1$ ponies that L has bitten. Each has been bitten by n other lions. This yields a count of $n(n+1)$ lions besides L. By Postulate 3, each lion other than L has bitten exactly one of these ponies, and is therefore counted exactly once. Hence there are exactly n^2+n+1 lions. In an analogous manner, we can prove that the total number of ponies is also n^2+n+1.

The community of lions and ponies was introduced by me in [3], with the ponies replaced by lambs. In thinking that the lambs were the most likely candidates to have been bitten, I totally missed the *point*.

Section 2. Starwars

Space Station Intelligentia received a call for help on the hyperradio from Spaceship Academia. Captain Philip said, "We are on the way home after a successful promotion of higher learning in distant star systems. We are surrounded by a Kleingon Fleet. Because we are on a peaceful mission, we are unarmed. Please send a relief force."

"Unfortunately, due to budget cuts, there are no other Spaceships on base at the moment," said Commander Gilbert. "Can you hold out?"

"Affirmative," said Captain Philip, "but we cannot disengage. It would help if you can get a Space Cannon to us."

"No problem. We will send one over by a Space Pod."

"Hang on a minute! Oh, no! There is a Space Tetropus in the Kleingon Fleet. It can grab one Space Pod at a time."

"I will send two Space Pods, each carrying a Space Cannon," said Commander Gilbert.

"Do not do that! Repeat! Do not do that!" Captain Philip said urgently. "If a Space Cannon falls into the hands of the Kleingons, we are history. It is too powerful even for us."

"I will send two, but only one has a Space Cannon. We will get the empty one to nudge up to the Space Tetropus."

"I don't think we can assume that the Space Tetropus is stupid."

"I wish I have one hundred Space Pods that I can send at the same time. With only one of them carrying a Space Cannon, our chance of success is 99%."

"That is easy for you to say, safely on the Space Station. Up here in the Spaceship, we do not like the 1% chance of failure."

"I will get back to you as soon as possible."

Commander Gilbert consulted Lieutenant Kenneth, the scientific advisor. He said, "We can break up a Space Cannon into two parts and send them separately. This way, the Kleingons can only get half of it, which is of absolutely no use to them."

"Unfortunately, Spaceship Academia will not get too much out of the other half either. However, your idea is an excellent one. If we break up two Space Cannons into two parts in identical fashion and send them by four Space Pods, the Kleingons will still be out of luck, while Spaceship Academia will have enough parts to reassemble a complete one."

The two officers were very pleased with their plan. However, when they tried to put it in operation, they found that there were only three Space Pods available on base.

Lieutenant Kenneth thought for a while and said, "Our main difficulty is not knowing which Space Pod the Space Tetropus will take. So to minimize our loss, we should distribute the Space Cannons as evenly as possible among the Space Pods."

"We must use two Space Cannons," said Commander Gilbert, "as we are bound to lose some parts. However, if we do not put the same part in the same Space Pod, we cannot lose them both. So two Space Cannons is exactly what we need to use."

"With two Space Cannons and three Space Pods, each Space Pod should carry two-thirds of a Space Cannon. So this means breaking up a Space Cannon into three parts, in identical fashion. Let us call them A, B and C. The first Space Pod will carry A and B, the second B and C, and the third C and A. So Spaceship Academia can still get a complete Space Cannon, while the Kleingons can only get two-thirds of one."

"It would be best if we do not break up the Space Cannons into too many parts. Couldn't we still have done it with only two?"

"No. Since we have four copies and three Space Pods, one of them must carry two. These must be different as there is no point in any Space Pod carrying two identical parts. If the Space Tetropus grabs this one, the Kleingons can put the two parts together to get a complete Space Cannon."

"I guess you are right," said Commander Gilbert. "It is lucky that we have three Space Pods. Had there been only two, we could not have done anything."

"Yes, each of Spaceship Academia and the Kleingons will get one. Either both have a chance of getting a complete Space Cannon, or neither has, which is definitely not good for us."

"Let us stop theorizing and put our plan to work. We cannot count on Spaceship Academia holding out forever against the Kleingons."

This was done, and soon words came over the hyperradio that all was well. Before long, Spaceship Academia was docking at Space Station Intelligentia. Commander Gilbert and Lieutenant Kenneth welcomed Captain Philip's safe return.

"That was a close call," reported Captain Philip. "The Kleingons were about to replace the Space Tetropus with a Space Octopus, which can grab two Space Pods at a time."

"This is serious," said Commander Gilbert. "Let us go to work at once and figure out a way around it, rather than wait until we have to face the situation."

"To begin with," said Lieutenant Kenneth, "we have to break up three Space Cannons. This way, we cannot lose every copy of any part. On the other hand, we do not need to break up more than three, as that will only make things easier for the Kleingons."

"Also, each Space Cannon must be broken up into at least three parts," Captain Philip said. "If there are only two, the Space Octopus can just nab one Space Pod carrying each part, and the Kleingons will have a complete Space Cannon. If we break it up into exactly three parts, we will need nine Space Pods so that each one will carry one part. Nothing less will do."

"We seldom have that many Space Pods on base," Commander Gilbert pointed out. "What is the smallest number of Space Pods that can carry out a successful convoy?"

"It has to be five or more. If we send only four, each side will get two, and that is bad news. It is the same argument which explains why two Space Pods are not enough for getting around a Space Tetropus."

"Are five Space Pods enough though?" Commander Gilbert pressed the point.

Nobody had an answer for a few days. Then Commander Gilbert found a diagram which his son Atticus drew.

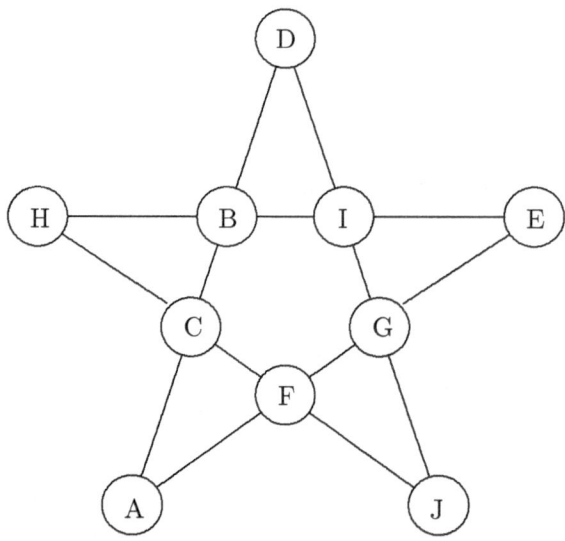

Figure 8.4

It was a regular five-pointed star, the emblem of Space Station Intelligentia. Atticus had labeled the ten points of intersections A, B, C, D, E, F, G, H, I and J in some random order, as shown in Figure 8.4.

"That gives me an idea," said Lieutenant Kenneth. "Let us break up each of three Space Cannons into ten parts labeled A to J. Let each of the five lines in Figure 8.4 represent a Space Pod, carrying all the parts whose labels do not appear on that line. Call DA line 1, AE line 2, EH line 3, HJ line 3 and JD line 5. Here is a list of the parts carried by the five Space Pods."

> Space Pod 1: E, F, G, H, I and J.
> Space Pod 2: B, C, D, H, I and J.
> Space Pod 3: A, C, D, F, G and J.
> Space Pod 4: A, B, D, E, G and I.
> Space Pod 5: A, B, C, E, F and H.

"Why would this work?" questioned Captain Philip.

"I see," said Commander Gilbert. "Whichever two Space Pods the Space Octopus captures, the two lines representing them will intersect. So it will be missing the part represented by that intersection. This is brilliant."

"Yes, indeed," agreed Captain Philip. "However, let us see if we can still work something out even if we have not been blessed with this divine revelation. Consider all possible scenarios. The Space Octopus may nab Space Pods 1 and 2, 1 and 3, 1 and 4, 1 and 5, 2 and 3, 2 and 4, 2 and 5, 3 and 4, 3 and 5, or 4 and 5. So for any of these ten pairs, there must be at least one part neither of which is carrying."

"Going back to what I said earlier," chimed in Lieutenant Kenneth, "we must have three copies of each part. Let the parts be labeled from A on. If Space Pods 1 and 2 are missing part A, then Space Pods 3, 4 and 5 must have it."

"This means that we cannot have two different pairs missing the same part," said Captain Gilbert, "even if the pairs overlap. In other words, we must break up each Space Cannon into ten parts, so that each of the ten pairs will be missing a different part."

"So 1 and 3 must be missing some part other than A, say B," said Captain Philip. "Then 2, 4 and 5 must have B. I think this will work. Let us draw a chart to show what each Space Pod should be carrying."

"Look!" exclaimed Lieutenant Kenneth. "You have come up with exactly the same solution obtained earlier!"

"What?" said Captain Philip. "Yes, I do believe you are right. It is nice to be able to get this in two ways."

Captured	1	1	1	1	2	2	2	3	3	4	S
Space Pods	2	3	4	5	3	4	5	4	5	5	P
Parts					E	F	G	H	I	J	1
carried		B	C	D				H	I	J	2
by each	A		C	D		F	G			J	3
of the	A	B		D	E		G		I		4
Space Pods	A	B	C		E	F		H			5

"I think you should write a book called *MENSA for Dummies* and put this in," laughed Commander Gilbert. "It is very pedantic and not brilliant at all, but it does work all the same."

The officers did not have much time to enjoy their discovery. Words just came in that the Kleingons had stepped up the arms race. Instead of replacing the Space Tetropus by a Space Octopus, they had replaced it by a Space Dodecopus. It had twelve arms and could grab three Space Pods at a time.

"Well," said Lieutenant Kenneth, "we must use four Space Cannons. We can break up each of them into only four parts if we have sixteen Space Pods."

"In view of the new development, we should have our budget increased. However, I seriously doubt that we could afford sixteen Space Pods. The minimum number of Space Pods we must have is seven," said Commander Gilbert.

"This would be a nightmare," complained Captain Philip. "Here is a list of thirty-five trios of seven Space Pods. This means we have to break up each Space Cannon into thirty-five parts. I am not sure if we can put them back together again."

(1,2,3)	(1,2,4)	(1,2,5)	(1,2,6)	(1,2,7)	(1,3,4)	(1,3,5)
(1,3,6)	(1,3,7)	(1,4,5)	(1,4,6)	(1,4,7)	(1,5,6)	(1,5,7)
(1,6,7)	(2,3,4)	(2,3,5)	(2,3,6)	(2,3,7)	(2,4,5)	(2,4,6)
(2,4,7)	(2,5,6)	(2,5,7)	(2,6,7)	(3,4,5)	(3,4,6)	(3,4,7)
(3,5,6)	(3,5,7)	(3,6,7)	(4,5,6)	(4,5,7)	(4,6,7)	(5,6,7)

The new budget allowed for eight Space Pods, one more than the absolute minimum.

"We have to break up each Space Cannon into at least fourteen parts," lamented Lieutenant Kenneth. "Suppose we have only thirteen parts. Then there are fifty-two pieces of equipments. On the average, each of the eight Space Pods must carry more than six pieces. Hence some Space Pod must carry at least seven parts."

"Let us assume that the Space Dodecopus will grab this one," said Commander Gilbert, picking up the train of thought. "Then the six parts it is missing are carried by the other seven Space Pods. Now there are twenty-four pieces of equipment. On the average, each of these seven Space Pods must carry more than three pieces, so that some Space Pod must carry at least four parts."

"Let this be the second Space Pod captured by the Space Dodecopus," said Captain Philip. "Now the two parts it is still missing are carried by the six Space Pods. Since there are eight pieces of equipment, some Space Pod must carry two pieces. If the Space Dodecopus captures this one too, it will have all the parts to put a Space Cannon back together.".

"Well, fourteen parts is much better than thirty-five parts," remarked Lieutenant Kenneth, "but can we do it with only fourteen parts?"

It was quite a while before any progress was made. The officers remembered the Finite Projective Geometry course they took during basic training. In particular, they recalled the Fano plane. They took two copies of it and labeled the points A, B, C, D, E, F and G in one, and T, U, V, W, X, Y and Z in the other, as shown in Figure 8.5.

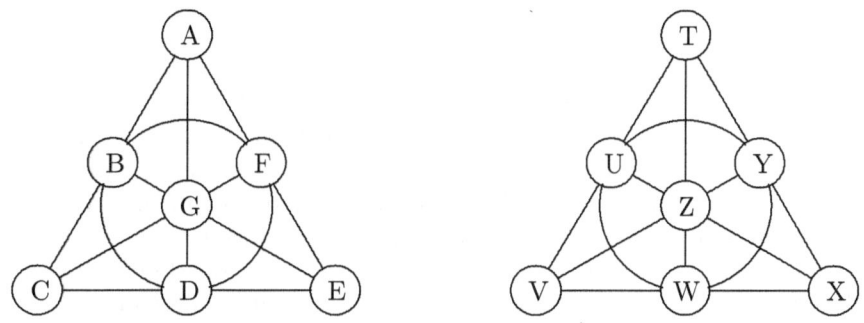

Figure 8.5

Captain Philip said, "Let line 1 be ABC, line 2 be CDE, line 3 be EFA, line 4 be ADG, line 5 by CFG, line 6 be BEG and line 7 be BDF. Following our earlier example, each line represents a Space Pod carrying all the parts whose labels do not appear on that line in the first copy. In addition, it carries all the parts whose labels do appear on that line in the second copy. The last Space Pod carries every part in the second copy. Here is a list of the parts carried by the eight Space Pods."

Space Pod 1: D, E, F, G, T, U and V.
Space Pod 2: A, B, F, G, V, W and X.
Space Pod 3: A, B, C, D, T, X and Y.
Space Pod 4: B, C, E, F, T, W and Z.
Space Pod 5: A, B, D, E, V, Y and Z.
Space Pod 6: A, C, D, F, U, X and Z.
Space Pod 7: A, C, E, G, U, W and Y.
Space Pod 8: T, U, V, W, X, Y and Z.

"Suppose the Space Dodecopus captures Space Pod 8. Then it captures only two of the other seven," said Lieutenant Kenneth. "The two lines representing them will intersect. So the Space Dodecopus will be missing the part represented by that intersection in the first copy."

"On the other hand," observed Commander Gilbert, "suppose that the Space Dodecopus does not capture Space Pod 8. In order for it to have all of T, U, V, W, X, Y and Z, they have to capture three Space Pods represented by lines all passing through the same point. However, the Space Dodecopus will be missing the part represented by that intersection in the first copy."

"So it works," said Captain Philip. "Let us hope that this stupid war ends soon."

Remark:
The problem in this section appears in [5] with a different story line. The results appeared in [2], a paper by Circle members Gilbert Lee, Kenneth Ng and Philip Stein. The extension into the Space Dodecopus are a small part of [1], a paper by Sven Chou and Jason Liao, two members of Chiu Chang Mathematical Circle. By the way, Kleingon was not a misspelling of Klingon of *Star Trek* fame. The word referred to an underling of a certain Albertan politician at the time, who was at odds with education.

Section 3. Convenient Buildings

A building is said to be *convenient* if for any two floors, there is at least one elevator which stops on both of them. Suppose the building has m elevators each of which stops on n floors. There are no restrictions on the choice of these floors. They do not have to be consecutive, and need not include the ground floor. What is the maximum number $f(m, n)$ of floors in this convenient building?

To establish the answer to this or any extremal problem, we need to do two things. First, we must show by an explicit construction that the answer can be attained. The finite projective planes would be useful for this purpose. Second, we must prove by a general argument that the answer cannot be improved.

We first prove three useful preliminary results.

Observation 1. $f(m + 1, n) \geq f(m, n)$.

Proof:
Having an extra elevator never hurts, though it may not help.

Observation 2. $f(m, n + 1) \geq f(m, n) + 1$.

Proof:
The extra stop for each elevator can all be on a new floor.

Observation 3. $f(m, kn) \geq kf(m, n)$.

Proof:
Pile k copies of a convenient building with $f(m, n)$ floors on top of one another to form a building with $kf(m, n)$ floors and connect the corresponding elevators in each copy so that each stops on kn floors. The same elevator which links the i-th and j-th floors in each copy will link the i-th floor of any copy to the j-th floor of any other copy. Thus the new building is convenient, and we have $f(m, kn) \geq kf(m, n)$.

We now study the function $f(m, n)$ by keeping m constant.

For $m = 1$, we have $f(1, n) = n$. The building can certainly have n floors. If it has more, the elevator will not stop on some floor. No elevator will stop on both this floor and any other floor.

For $m = 2$, we still have $f(2, n) = n$. By Observation 1, $f(2, n) \geq f(1, n) = n$. If the building has more floors, each elevator will not stop on some floor. If they skip different floors, no elevator will stop on both. If they skip the same floor, no elevator will stop on both this floor and any other floor.

The first interesting case is $m = 3$. Let there be three floors 1, 2 and 3. Let the first elevator stop on floors 1 and 2, the second elevator on floors 1 and 3, and the third elevator on 2 and 3. Thus we have the lower bound $f(3, 2) \geq 3$. This is a *perfect* building because there is no duplication of services.

We next prove that $f(3, 2k) = 3k$. By $f(3, 2) \geq 3$ and Observation 3, $f(3, 2k) \geq 3k$. The total number of stops is $6k$. If each floor is served by at least 2 elevators, then the number of floors is at most $3k$. If some floor is served by at most 1 elevator, it can be linked to at most $2k - 1$ other floors. Counting this floor, the building can have at most $2k$ floors. It follows that $f(3, 2k) = 3k$.

We now prove that $f(3, 2k + 1) = 3k + 1$. By Observation 2,

$$f(3, 2k + 1) \geq f(3, 2k) + 1 = 3k + 1.$$

The total number of stops is $6k + 3$. If each floor is served by at least 2 elevators, then the number of floors is at most $3k + 1$. If some floor is served by at most 1 elevator, it can be linked to at most $2k$ other floors. Counting this floor, the building can have at most $2k + 1$ floors. It follows that $f(3, 2k + 1) = 3k + 1$.

The cases $m = 4$ and $m = 5$ are slightly more difficult because of the absence of perfect buildings. The next perfect building occurs at $m = 6$. Here the floors are 1, 2, 3 and 4, and each of the six elevators stop on a different pair of the four floors, namely, (1,2), (1,3), (1,4), (2,3), (2,4) and (3,4). This yields the lower bound $f(6, 2) \geq 4$.

We next prove that $f(6, 2k) = 4k$. By $f(6, 2) \geq 4$ and Observation 3, $f(6, 2k) \geq 4k$. The total number of stops is $12k$. If each floor is served by at least 3 elevators, then the number of floors is at most $4k$. If some floor is served by at most 2 elevators, it can be linked to at most $4k - 2$ other floors. Counting this floor, the building can have at most $4k - 1$ floors. It follows that $f(6, 2k) = 4k$.

We now prove that $f(6, 2k+1) \leq 4k+2$. Observe that the total number of stops is $12k + 6$. If each floor is served by at least 3 elevators, then the number of floors is at most $4k + 2$. If some floor is served by at most 2 elevators, it can be linked to at most $4k$ other floors. Counting this floor, the building can have at most $4k + 1$ floors.

Finally, we give a general construction to show that $f(6, 2k+1) \geq 4k+2$. Let the floors be $a_1, a_2, \ldots, a_k, \ b_1, b_2, \ldots, b_k, \ c_1, c_2, \ \ldots, c_k, \ d_1, d_2, \ldots, d_k, \ e$ and f. Let the first elevator stop at all the a's and b's, the second at all the a's and c's, the third at all the a's and d's, the fourth at all the b's and c's, the fifth at all the b's and d's, and the sixth at all the c's and d's. Then these $4k$ floors are all linked.

If we add e as the last stop of the first and sixth elevator and f as the last stop of the second and fifth elevator, they are also linked to the other $4k$ floors. However, e and f are not linked. So we replace d_k by f in the sixth elevator. This destroys the links between d_k on the one hand and e and the c's on the other. The remedy is to add e as the last stop of the third elevator and d_k as the last stop of the fourth elevator. It follows that $f(6, 2k+1) = 4k+2$.

The Fano plane is the next example of a perfect building. It leads to the lower bound $f(7,3) \geq 7$. By this and Observation 3, we have $f(7, 3k) \geq 7k$. Now the total number of stops is $21k$. If each floor is served by at least 3 elevators, then the number of floors is at most $7k$. If some floor is served by at most 2 elevators, it can be linked to at most $6k - 2$ other floors. Counting this floor, the building can have at most $6k - 1$ floors. It follows that $f(7, 3k) = 7k$.

Our final result is that $f(7, 3k+2) = 7k+4$. To prove that $f(7, 3k+2) \leq 7k + 4$, observe that the total number of stops is $21k + 14$. If each floor is served by at least 3 elevators, then the number of floors is at most $7k + 4$. If some floor is served by at most 2 elevators, it can be linked to at most $6k+2$ other floors. Counting this floor, the building can have at most $6k+3$ floors.

We now give a general construction to show that $f(7, 3k + 2) \geq 7k + 4$. Let the floors be a_1, a_2, $\ldots$, a_{k+1}, b_1, b_2, $\ldots$, b_{k+1}, c_1, c_2, $\ldots$, c_{k+1}, d_1, d_2, $\ldots$, d_{k+1}, e_1, e_2, $\ldots$, e_k, f_1, f_2, $\ldots$, f_k and g_1, g_2, $\ldots$, g_k. Let the first elevator stop at all the a's, b's and e's, the second at all the a's, c's and f's, the third at all the a's, d's and g's, the fourth at all the b's, c's and g's, the fifth at all the b's, d's and f's, the sixth at all the c's, d's and e's, and the seventh at all the e's, f's and g's. Then all the floors are linked, with two wasted stops in the seventh elevator. It follows that $f(7, 3k + 2) = 7k + 4$.

We are unable to determine $f(7, 3k + 1)$.

The results in this section are contained in [4], the work of Jerry Lo, a member of Chiu Chang Mathematical Circle, and Circle David Rhee.

Exercises

1. Prove the duals of the four postulates of the community of lions and ponies:

   (a) There are at least two ponies.

   (b) Each pony has been bitten by at least three lions.

   (c) For any pair of ponies, there is exactly one lion that has bitten both.

   (d) For any pair of lions, there is at least one pony that both have bitten.

2. Show that to get around the Space Dodecapus with nine Space Pods, it is enough to break up each of four Space Cannons into twelve parts.

3. Prove that $7k+1 \le f(7, 3k+1) \le 7k+2$ where $f(m, n)$ is the function associated with convenient buildings.

Bibliography

[1] Sven Chou and Jason Liao, Reliable Delivery with Unreliable Delivers, *Delta-K* **46**-1 (2008) 33–36.

[2] Gilbert Lee, Kenneth Ng and Philip Stein, A Space Interlude, in *Mathematics for Gifted Students II*, a special edition of *Delta-K*, **33**-3 (1996) 31–32.

[3] Andy Liu, Lions, Lambs and Proofs, Math. Inform. Quart. **10** (2000) 131–132.

[4] Jerry Lo and David Rhee, The Elevator Problem, in "Mathematical Wizardry for a Gardner", edited by Ed Pegg Jr., Alan H. Schoen and Tom Rodgers, A K Peters, Natick (2009) 165–171.

[5] Dennis Shasha, *The Puzzling Adventures of Dr. Ecco*, Dover Publications Inc., Mineola (1998) 115–118 and 173–175.

Chapter Nine: Sharing Loots

Section 1. Sharing Jewels.

Captain Crook and his pirates had salvaged from the Spanish Main a giant necklace. It consisted of six enormous jewels linked together by five chains, each connecting one jewel to another. The value of each jewel was assessed and announced to everyone.

It was agreed that Captain Crook would keep three jewels for himself, while the crew would share the proceed from selling the other three. Captain Crook and the crew would alternately take one jewel at a time, cutting one chain which would separate the jewel being taken from the rest. Captain Crook would start, and when he made the last cut which would separate the last two jewels, he could choose either one.

With the advantage of going first, Captain Crook would not be satisfied unless he could get three jewels whose total worth was at least as much as the total worth of the other three. In that case, he was said to have won. Now there were many different configurations for the necklace. For the same configuration, jewels of different values might be in different locations. Could Captain Crook always win?

An exhaustive analysis appears daunting. So let us conduct some preliminary analysis. A basic technique in problem-solving is *down-sizing*. Instead of dealing with six jewels, we reduce their number while keeping it even.

Figure 9.1

Figure 9.1 shows a necklace with two jewels, where the black circles represent the jewels and the line segment represents the chain. Since there is only one chain, the two jewels will be separated once Captain Crook cuts the chain. Obviously, he will take the jewel of higher value. (If two jewels have equal value, either may be considered to have higher value). In this case, Captain Crook can win.

Figure 9.2

When there are four jewels, there are two different ways in which they may be linked together. In the first case, as shown in Figure 9.2, there are three jewels which Captain Crook can take in his opening move.

© Springer International Publishing AG 2018

A. Liu, *S.M.A.R.T. Circle Projects*, Springer Texts in Education, DOI 10.1007/978-3-319-56811-9_9

In the opening round, each side takes one of these three jewels. Clearly, Captain Crook will take the available jewel of the highest value. Thus he will get at least as much value as the crew in this round. Now there are only two jewels left, and we already know that Captain Crook can get at least half of the total value of the remaining jewels. It follows that Captain Crook can win.

For later references, we give this idea a formal name.

The Induction Argument.
Suppose we have proved that Captain Crook wins when the number of jewels is $2n$ for some positive integer n. Consider the case when the number of jewels is $2n+2$. If Captain Crook can get at least as much value as the crew in the opening round, then he can win.

Figure 9.3

In the second case, as shown in Figure 9.3, there are two jewels available initially, but it may be counter-productive for Captain Crook to take the one of higher value. The reason is that this action may made available a previously inaccessible jewel that may be more valuable. We need a new idea.

Captain Crook repaints two of the jewels white as shown in Figure 9.3. If the total value of the two black jewels is at least as much as the total value of the two white jewels, he takes the available black jewel. The crew have no choice but to take a white jewel, which allows Captain Crook to take the other black jewel and win. The situation is symmetric if the total value of the two white jewels is more than the total value of the two black jewels.

Again, for later references, we give this idea a formal name.

The Coloring Argument.
Suppose Captain Crook can paint the $2n$ jewels with n in each of two color, and can take all the jewels of either color, then he wins.

Figure 9.4

Having solved the simpler problems with two and four jewels, we now return to the original problem with six jewels. There are six different configurations for the necklace. In the first two cases, as shown in Figure 9.4, Captain Crook wins by induction.

Figure 9.5

In the third case, as shown in Figure 9.5, Captain Crook wins by coloring.

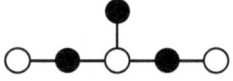

Figure 9.6

In the fourth case, as shown in Figure 9.6, Captain Crook checks to see whether the total value of the black jewels is at least as much as the value of the white jewels. If so, he takes the only black jewel that is available and wins by coloring. If not, he takes the available white jewel with of higher value. If the crew take a black jewel, Captain Crook wins by coloring. If the crew take the white jewel, Captain Crook wins by induction.

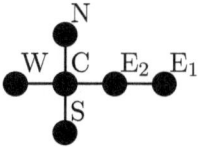

Figure 9.7

Figure 9.7 shows the fifth case, with a label for each of the jewels. Suppose E_1 has no less value than E_2. Captain Crook takes the available jewel with the most value and wins by induction. Suppose E_2 has higher value than E_1. Captain Crook takes among N, W and S the one with the most value. If the crew also take from what are left of N, W and S, Captain Crook wins by induction. Hence the crew must take E_1. Now Captain Crook takes E_2. Since the crew must now take from what are left of N, W and S, Captain Crook wins by a modified induction argument, getting higher value of the opening two rounds instead of in the opening round.

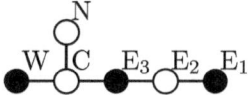

Figure 9.8

Figure 9.8 shows the sixth case with a label for each of the the jewels. Suppose E_1 has no less value than E_2. Captain Crook takes the available jewel with the most value and wins by induction. Suppose E_2 has higher value than E_1. Captain Crook paints in white C, E_2 and whichever of N and W has higher value, say N. Then he checks whether the total value of the white jewels is at least as much as the total value of the black jewels. If so, he takes N and wins by coloring. If not, Captain Crook takes E_1. If the crew now take N, Captain Crook wins by coloring. If the crew instead take W, Captain Crook takes N. Although he does not get all the black jewels, he still wins by a modified coloring argument, getting a jewel (N) in the minority color with higher value of place of a jewel (W) in the majority color with less value. Finally, if the crew take E_2, Captain Crook takes E_3. If the crew then take N, Captain Crook wins by coloring. If the crew instead take W, Captain Crook takes N and wins by modified coloring. If C has higher value than N, of course Captain Crook will take C instead, and be still better off.

This work [5] by Hsin-Po Wang, a member of Chiu Chang Mathematical Circle, was motivated by the "Coins in a row" problem in [3]. For other intriguing problems, see [4] by the same author.

Section 2. Sharing Gold and Silver.

Captain Crook and his pirates had salvaged 100 boxes from the Spanish Main. Each contained some gold and some silver. Naturally, he wanted the lion's share, after which his crew would divide the rest among themselves. In order not to appear overly greedy, he limited himself to taking only a certain number of boxes. However, he had to make this announcement right away, and there was no time to examine in detail the contents of each box. Of course, he would have the chance to do so when he made his selections. What was the minimum number of boxes he must take in order to guarantee that he could get at least one half of the total amount of gold and one half the total amount of silver?

General Formulation.

In each of m boxes, there are k different metals in varying amounts. Suppose we wish to get at least $\frac{p}{q}$ of the total amount of each metal. We wish to determine $f(k, \frac{p}{q}, m)$, the minimum number of boxes we must take regardless of the distribution of metals among the boxes.

We will focus on $k = 2$. We begin our investigation with a simple general observation.

Lemma 1.

We have $f(k, \frac{p}{q}, m) \le f(k, \frac{p}{q}, m+1) \le f(k, \frac{p}{q}, m) + 1$.

Proof:

To establish the lower bound, consider the special case where one box is empty. Set it aside, and we have to take $f(k, \frac{p}{q}, m)$ of the remaining ones in order to get $\frac{p}{q}$ of each metal. To establish the upper bound, we set aside an arbitrary box, and take $f(k, \frac{p}{q}, m)$ of the remaining one so as to get at least $\frac{p}{q}$ of each metal in these m boxes. Adding the box set aside to our collection will not reduce this fraction.

The cases $p = 1$ and $p = q - 1$ yield most readily to our approach. For $q = 2$, these two cases merge into a single case $p = 1$, which is the original problem when $k = 2$.

Theorem 1A.

For $q(n-1) + 1 \le m \le qn$, $f(1, \frac{1}{q}, m) = n$.

Proof:

To establish the lower bound, we consider the distribution with the same amount of gold in each box. Since there are at least $q(n-1) + 1$ equal shares, we must take n boxes. To establish the upper bound, consider any distribution and arrange the boxes in non-ascending order of amount of gold. If we take the first box, skip the next $q-1$, take the next box, skip the next $q-1$, and so on, we would have taken n boxes with at least $\frac{1}{q}$ of the total amount of gold.

Theorem 1B.

For $qn \leq m \leq qn + q - 1$, $f(1, \frac{q-1}{q}, m) = m - n$.

Proof:

To establish the lower bound, we consider the distribution with the same amount of gold in each box. To get at least $\frac{q-1}{q}$ of the gold, the number of boxes we must take is at least

$$\left\lceil \frac{(q-1)m}{q} \right\rceil = m - \left\lfloor \frac{m}{q} \right\rfloor = m - n.$$

To establish the upper bound, consider any distribution and arrange the boxes in non-ascending order of amount of gold. If we take the first $q - 1$ boxes, skip the next, take the next $q - 1$ boxes, skip the next, and so on. we would have taken $m - n$ boxes with at least $\frac{1}{q}$ of the total amount of gold.

The next observation is the key in going from $k = 1$ to $k = 2$.

Lemma 2.

Each of m boxes contains some gold. If the maximum amount of gold in any of the boxes is a kilograms, then the boxes can be divided into q groups such that the total number of boxes in one group differs from that in any other by at most 1, and the total amount of gold in the boxes of one group differs from that of any other by at most a kilograms.

Proof:

Let $m = qn + i$ where $0 \leq i < q$. Let the amount of gold in the t-th box be a_t kilograms. We may assume that $a = a_1 \geq a_2 \geq \cdots \geq a_m$. Put all the boxes with indices congruent to ℓ (mod q) in the ℓ-th group for $i > 0$ and all the boxes with indices divisible by q in the q-th group. Clearly the total number of boxes in one group differs from that in any other by at most 1, the total amount of gold in the boxes of the first group is the largest, and the total amount of gold in the boxes of the last group is the smallest, but the difference is $a_1 - (a_q - a_{q+1}) - \cdots - (a_{qn} - a_{qn+1}) \leq a_1 = a$ kilograms.

Theorem 2A.

For $q(n-1) + 2 \leq m \leq qn + 1$, $f(2, \frac{1}{q}, m) = n + 1$.

Proof:

We first give a specific construction which establishes the lower bound. Suppose the gold is all in one box while the silver is evenly distributed among the other boxes. Then we must take the box containing all the gold. Since there are at least $q(n-1) + 1$ equal shares of silver, we must take n of the remaining boxes. To establish the upper bound, set aside the box with the highest amount of gold. Let this amount be a_0 kilograms, and let the highest amount of gold in any of the other boxes be a_1 kilograms.

By Lemma 2, we may divide the remaining boxes into q groups such that the total number of boxes in one group differs from the total number of boxes in any other by at most 1, and the total amount of gold in the boxes of one group differs from the total amount of gold in the boxes of the other group by at most a_1 kilograms. Now choose the group of boxes containing the highest total amount of silver, and take as well the box set aside initially. We have taken either n or $n+1$ boxes. Clearly, we have taken at least $\frac{1}{q}$ of the silver. Now the group of boxes taken contains a total amount of gold at most a_1 kilograms less than the total amount of gold in the boxes of any other group. Since $a_0 \geq a_1$, adding a_0 kilograms means that we have taken at least $\frac{1}{q}$ of the gold.

Theorem 2B.

For $qn - 1 \leq m \leq qn + q - 2$, $f(2, \frac{q-1}{q}, m) = m - n + 1$.

Proof:

We first give a specific construction which establishes the lower bound. Suppose the gold is evenly distributed among $q - 1$ boxes and the silver among the other boxes. Then we must take the $q - 1$ boxes containing all the gold and at least $\lceil \frac{(q-1)(m-q+1)}{q} \rceil$ of the remaining boxes. The total is $q - 1 + m - q + 1 - \lfloor \frac{m-q+1}{q} \rfloor = m - n + 1$. To establish the upper bound, set aside the $q - 1$ boxes with the highest amounts of gold. Let the highest amount of gold in any of the other boxes be a kilograms. By Lemma 2, we may divide the remaining boxes into q groups such that the total number of boxes in one group differs from the total number of boxes in any other by at most 1, and the total amount of gold in the boxes of one group differs from the total amount of gold in the boxes of the other group by at most a kilograms. Now choose the $q - 1$ groups of boxes containing the highest total amounts of silver. Take also the $q - 1$ boxes set aside initially, each containing at least a kilograms of gold. Clearly, we have at least $\frac{q-1}{q}$ of the silver and at least $\frac{q-1}{q}$ of the gold. Since the group of boxes we skip contains n or $n - 1$ boxes, we have taken at most $m - n + 1$ boxes.

When $1 < p < q - 1$, the problem becomes more difficult except when $k = 1$. Theorems 1A and 1B are corollaries of the following result.

Theorem 3.

Write $m = qn + i$ where $0 \leq i < q$. Then $f(1, \frac{p}{q}, qn + i) = pn + j$ where $j = \lceil \frac{ip}{q} \rceil$.

Proof:

To establish the lower bound, we consider the distrubtion with the same amount of gold in each box, say 1 kilogram. Then $\frac{p}{q}$ of the total amount is $\frac{p}{q}(qn + i) = pn + \frac{ip}{q}$ kilograms. Hence we must take $pn + j$ boxes.

To establish the upper bound, consider any distribution and arrange the boxes in non-ascending order of amount of gold. Divide them into n groups of q, with i boxes left over. If we take the first p boxes of each group and the first j boxes of the left over, we certainly have at least $\frac{p}{q}$ of the total amount of gold.

Lemma 3.
Each of several boxes contains some gold and some silver. They are arranged in non-ascending order of the amounts of gold inside them. In problems where we try to get at least a certain amount of gold and at least a certain amount of silver by taking a certain number of boxes, we may assume that the amounts of silver inside the boxes are in non-descending order.

Proof:
Suppose there exist two boxes X and Y such that there are more gold and more silver in X than in Y. Simply interchange their silver contents. Suppose the boxes are now in non-descending order of the amount of silver inside, and we proceed to choose our boxes. If this calls for the taking of both or neither of the new boxes, we may pretend that the interchange never happened. If this calls for the taking of either one of the new boxes but not the other, we can always do better by taking X. Suppose the interchange does not produce immediately the desired arrangement. Then further interchanges can be made. Moreover, only a finite number of interchanges are needed to obtain the desired arrangement.

The first meaningful case for $k = 2$ and $1 < p < q - 1$ is when $q = 5$. We now show how instrumental Lemma 3 is in the analysis.

Theorem 4A.
We have

$$f(2, \tfrac{2}{5}, 5n) \quad\;\; = \;\; 2n + 1;$$

$$f(2, \tfrac{2}{5}, 5n + 1) \;\; = \;\; 2n + 2;$$

$$f(2, \tfrac{2}{5}, 5n + 2) \;\; = \;\; 2n + 2;$$

$$f(2, \tfrac{2}{5}, 5n + 3) \;\; = \;\; 2n + 2$$

$$\text{and}\quad f(2, \tfrac{2}{5}, 5n + 4) \;\; = \;\; 2n + 3.$$

Proof:
We first give a specific construction which establishes the lower bound. Suppose the gold is evenly distributed among three boxes while the silver is evenly distributed among the other boxes. Then we must take two of the boxes containing gold.

We now consider the following two cases:

Case 1. $f(2, \frac{2}{5}, 5n + 4) \geq 2n + 3$.

Since $\frac{2n}{5n+1} < \frac{2}{5} < \frac{2n+1}{5n+1}$, we must take $2n + 1$ of the $5n + 1$ boxes containing silver.

Case 2. $f(2, \frac{2}{5}, 5n + 2) \geq 2n + 2$.

Since $\frac{2n-1}{5n-2} < \frac{2}{5} < \frac{2n}{5n-2}$, we must take $2n$ of the $5n - 2$ boxes containing silver.

By Lemma 1, $f(2, \frac{2}{5}, 5(n + 1)) \geq f(2, \frac{2}{5}, 5n + 4) \geq 2(n + 1) + 1$. Hence $f(2, \frac{2}{5}, 5n) \geq 2n + 1$. By Lemma 1 again,

$$f(2, \frac{2}{5}, 5n + 3) \geq f(2, \frac{2}{5}, 5n + 2) \geq f(2, \frac{2}{5}, 5n + 1) \geq 2n + 2.$$

To establish the upper bound, assume that the boxes have been arranged according to Lemma 3. We consider the following two cases:

Case 1. $f(2, \frac{2}{5}, 5n) \leq 2n + 1$.

Divide the boxes into groups of five. In each group, take the first and third boxes. Also take the very last box. We have $\frac{a_1+a_3}{a_1+a_2+a_3+a_4+a_5} \leq \frac{2}{5}$ since this is equivalent to $3(a_1 + a_3) \geq 2(a_2 + a_4 + a_5)$, which holds since $a_1 \geq a_2 \geq a_3 \geq a_4 \geq a_5$. Thus we have taken at least $\frac{2}{5}$ of the gold in the first group. Since the situation is repeated in the other groups, and taking the last box can only help, we have at least $\frac{2}{5}$ of all the gold. For silver, start from the other end and switch the last box of each group with the first box of the next group. The amount of silver in the boxes of each group is still in non-ascending order, and we have exactly the same situation as with gold. Figure 9.9 illustrates the case $n = 3$.

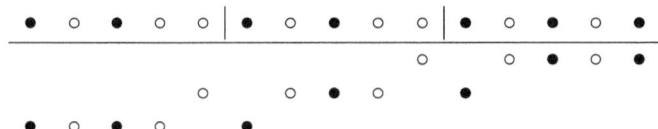

Figure 9.9

Case 2. $f(2, \frac{2}{5}, 5n + 3) \leq 2n + 2$.

Divide the boxes into groups of five, with the last group consisting only of three boxes. In each group including the last incomplete group, take the first and third boxes. As in Case 1, we have taken at least $\frac{2}{5}$ of all the gold. For silver, start from the other end. This time, we already have exactly the same situation as with gold. Figure 9.10 illustrates the case $n = 3$.

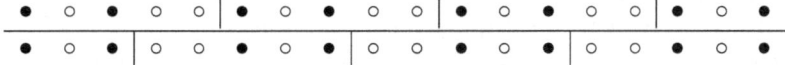

Figure 9.10

By Lemma 1, $f(2, \frac{2}{5}, 5n+4) \leq f(2, \frac{2}{5}, 5(n+1)) \leq 2(n+1)+1 = 2n+3$. By Lemma 1 again, $f(2, \frac{2}{5}, 5n+1) \leq f(2, \frac{2}{5}, 5n+2) \leq f(2, \frac{2}{5}, 5n+3) \leq 2n+2$.

Theorem 4B.

We have

$$f\left(2, \frac{3}{5}, 5n\right) \qquad = \quad 3n+1;$$

$$f\left(2, \frac{3}{5}, 5n+1\right) \quad = \quad 3n+2;$$

$$f\left(2, \frac{3}{5}, 5n+2\right) \quad = \quad 3n+2;$$

$$f\left(2, \frac{3}{5}, 5n+3\right) \quad = \quad 3n+3$$

$$\text{and} \quad f\left(2, \frac{3}{5}, 5n+4\right) \quad = \quad 3n+4.$$

Proof:

We first give a specific construction which establishes the lower bound. Suppose the gold is evenly distributed among two boxes while the silver is evenly distributed among the other boxes. Then we must take both boxes containing all the gold. We now consider the following three cases:

Case 1. $f(2, \frac{3}{5}, 5n+4) \geq 3n+4$.

Since $\frac{3n+1}{5n+2} < \frac{3}{5} < \frac{3n+2}{5n+2}$, we must take $3n+2$ of the $5n+2$ boxes containing silver.

Case 2. $f(2, \frac{3}{5}, 5n+1) \geq 3n+2$.

Since $\frac{3n-1}{5n-1} < \frac{3}{5} < \frac{3n}{5n-1}$, we must take $3n$ of the $5n-1$ boxes containing silver.

Case 3. $f(2, \frac{3}{5}, 5n+3) \geq 3n+3$.

Since $\frac{3n}{5n+1} < \frac{3}{5} < \frac{3n+1}{5n+1}$, we must take $3n+1$ of the $5n+1$ boxes containing silver.

By Lemma 1, $f(2, \frac{3}{5}, 5(n+1)) \geq f(2, \frac{3}{5}, 5n+4) \geq 3(n+1)+1$. Hence $f(2, \frac{3}{5}, 5n) \geq 3n+1$. By Lemma 1 again,

$$f\left(2, \frac{3}{5}, 5n+2\right) \geq f\left(2, \frac{3}{5}, 5n+1\right) \geq 3n+2.$$

To establish the upper bound, assume that the boxes have been arranged according to Lemma 3. We consider the following two cases:

Case 1. $f(2, \frac{3}{5}, 5n) \leq 3n+1$.

Divide the boxes into groups of five. In each group, take the first, second and fourth boxes. Also take the very last box. We have $\frac{a_1+a_2+a_4}{a_1+a_2+a_3+a_4+a_5} \leq \frac{3}{5}$ since this is equivalent to $2(a_1 + a_2 + a_4) \geq 3(a_3 + a_5)$, which holds since $a_1 \geq a_2 \geq a_3 \geq a_4 \geq a_5$. Thus we have taken at least $\frac{3}{5}$ of the gold in the first group. Since the situation is repeated in the other groups, and taking the last box can only help, we have at least $\frac{3}{5}$ of all the gold.

For silver, start from the other end and switch the last box of each group with the first box of the next group. The amount of silver in the boxes of each group is still in non-ascending order, and we have exactly the same situation as with gold. Figure 9.11 illustrates the case $n = 3$.

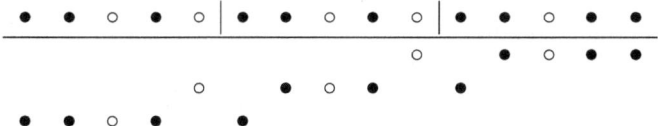

Figure 9.11

Case 2. $f(2, \frac{3}{5}, 5n + 2) \leq 3n + 2$.
Divide the boxes into groups of five, with the last group consisting only of two boxes. In each group, take the first, second and fourth boxes. For the last incomplete group, this means taking both boxes. As in Case 1, we have taken at least $\frac{3}{5}$ of all the gold. For silver, start from the other end. This time, we already have exactly the same situation as with gold. Figure 9.12 illustrates the case $n = 3$.

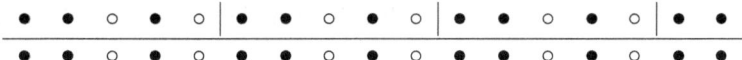

Figure 9.12

By Lemma 1,

$$
\begin{array}{ccccc}
f(2, \frac{3}{5}, 5n + 1) & \leq & f(2, \frac{3}{5}, 5n + 2) & \leq & 3n + 2, \\
f(2, \frac{3}{5}, 5n + 3) & \leq & f(2, \frac{3}{5}, 5n + 2) + 1 & \leq & 3n + 3, \\
f(2, \frac{3}{5}, 5n + 4) & \leq & f(2, \frac{3}{5}, 5n + 3) + 1 & \leq & 3n + 4.
\end{array}
$$

At this point, we made a bold conjecture. We claim that

$$
\boxed{f\left(k, \frac{p}{q}, qn + i\right) = pn + j,}
$$

where $0 \leq i \leq q - 1$ and $j = (k - 1) + \lceil \frac{pi - (k-1)}{q} \rceil$.

This general formula agrees with all preceding results. We now show that it is at least a lower bound.

Theorem 5.

If $0 \le i \le q - 1$ and $j = (k - 1) + \lceil \frac{pi - (k-1)}{q} \rceil$, then $f(k, \frac{p}{q}, qn + i) \ge pn + j$,

Proof:

Let $\frac{r}{s}$ be the lower Farey fraction which generates $\frac{p}{q}$. Then $0 < s < q$, $\frac{r}{s} < \frac{p}{q} < \frac{r+1}{s}$ and $ps = qr + 1$. For each metal except the last one, distribute the total amount equally among s boxes which contain no other metals. From each group, we must take $r + 1$ boxes for a total of $(k - 1)(r + 1)$. Distribute the total amount of the last metal equally among the remaining $qn + i - (k - 1)s$ boxes, and we must take from this group $\lceil \frac{p(qn+i) - (k-1)s}{q} \rceil$ boxes. For $0 \le i \le q - 1$,

$$
\begin{aligned}
f\left(k, \frac{p}{q}, qn + i\right) &\ge (k - 1)(r + 1) + \left\lceil \frac{p(qn + i) - (k - 1)s}{q} \right\rceil \\
&= (k - 1)(r + 1) + pn + \left\lceil \frac{pi - (k - 1)(qr + 1)}{q} \right\rceil \\
&= pn + (k - 1) + \left\lceil \frac{pi - (k - 1)}{q} \right\rceil \\
&= pn + j,
\end{aligned}
$$

where $j = (k - 1) + \lceil \frac{pi - (k-1)}{q} \rceil$.

We are as yet unable to establish this formula as an upper bound.

This problem is a special case of a general problem posed in the 2005 International Mathematics Tournament of the Towns Summer Seminar in Mir Town, Belarus, attended by Circle member David Rhee. For results on $q = 7$, 8 and 9, see [2].

Section 3. Sharing Rum.

Captain Crook and his pirates had salvaged a giantic barrel of rum from the Spanish Main. They would share it with the aid of four empty barrels. In the first stage, the rum would be distributed among the five barrels. The crew and Captain Crook alternately subdivide the content of a barrel into two, pouring some into an empty barrel. The amount poured could be 0; in other words, either side could pass. The crew would make the first move so that Captain Crook would get the last move.

In the second stage, Captain Crook and the crew would choose the barrels alternately. Captain Crook chose first, thereby getting three barrels while the crew would get only two. What was the maximum amount of rum he could get?

The total volume of the rum is taken to be 1. By the volume of a barrel, we mean the volume of the rum in that barrel. A barrel is said to be larger than another barrel if the first one contains more rum than the second.

Suppose in the first move. the crew subdivides it into $a \leq b$. There are four cases.

Case A. $\frac{34}{53} \leq a \leq 1$ so that $0 \leq b \leq \frac{19}{53}$.
Captain Crook subdivides b into $\frac{b}{2}$ and $\frac{b}{2}$. If the crew does not subdivide a, Captain Crook just pours out $\frac{1}{53}$ from another barrel. He will get three barrels with total volume at least $a + \frac{1}{53} \geq \frac{35}{53}$. Suppose the crew subdivides a into $c \leq d$. Captain Crook subdivides d into $\frac{d}{2}$ and $\frac{d}{2}$. He will get three barrels with total volume at least $c + \frac{d}{2} + \frac{b}{2} \geq \frac{a}{4} + \frac{1}{2} \geq \frac{35}{53}$.

Case B. $\frac{33}{53} \leq a \leq \frac{34}{53}$ so that $\frac{19}{53} \leq b \leq \frac{20}{53}$.
Captain Crook subdivides b into $\frac{18}{53}$ and $b - \frac{18}{53}$. If the crew does not subdivide a, Captain Crook just pours out $\frac{2}{53}$ from another barrel. He will get three barrels with total volume at least $a + \frac{2}{53} \geq \frac{35}{53}$. Suppose the crew subdivides a into $c \leq d$. We consider four subcases.

Subcase B1. $\frac{18}{53} \geq c \geq \frac{17}{53} \geq d \geq b - \frac{18}{53}$.
Captain Crook subdivides c into d and $c - d$. He will get three barrels with total volume at least $\frac{18}{53} + d + \min\{c - d, b - \frac{18}{53}\}$. In the former instance, it is at least $\frac{18}{53} + c \geq \frac{35}{53}$. In the latter instance, it is at least $d + b = 1 - c \geq \frac{35}{53}$.

Subcase B2. $\frac{17}{53} \geq c \geq d \geq b - \frac{18}{53}$.
Captain Crook subdivides $b - \frac{18}{53}$ into $\frac{b}{2} - \frac{9}{53}$ and $\frac{b}{2} - \frac{9}{53}$. He will get three barrels with total volume at least $\frac{18}{35} + d + \frac{b}{2} - \frac{9}{35} = \frac{62}{53} - \frac{b}{2} - c \geq \frac{35}{53}$.

Subcase B3. $c \geq \frac{18}{53} \geq d \geq b - \frac{18}{53}$.
Captain Crook subdivides $\frac{18}{53}$ into d and $\frac{18}{53} - d$. He will get three barrels with total volume at least $c + d + \min\{\frac{18}{35} - d, b - \frac{18}{35}\}$. In the former instance, it is at least $c + \frac{18}{53} \geq \frac{36}{53}$. In the latter instance, it is at least $a + b - \frac{18}{53} = \frac{35}{53}$.

Subcase B4. $c \geq \frac{18}{53} \geq b - \frac{18}{53} \geq d$.

Captain Crook subdivides $\frac{18}{53}$ into $\frac{9}{53}$ and $\frac{9}{53}$. Since $d \leq b - \frac{18}{53} \leq \frac{2}{53}$, $c \geq \frac{31}{53}$. Hence he will get three barrels with total volume at least $c + \frac{9}{53} \geq \frac{40}{53}$.

Case C. $\frac{27}{53} \leq a \leq \frac{33}{53}$ so that $\frac{20}{53} \leq b \leq \frac{26}{53}$.

Catpain Corrk subdivides a into $\frac{27}{53}$ and $a - \frac{27}{53}$. If the crew does not subdivide $\frac{27}{53}$, Captain Crook just pour out $\frac{8}{53}$ from another barrel. If it is the second largest, then the crew gets two barrels with total volume at most $\frac{16}{53}$. Otherwise, Captain Crook will get three barrels with total volume at least $\frac{27}{53} + \frac{8}{53} = \frac{35}{53}$. Suppose the crew subdivides $\frac{27}{53}$ into $c \geq d$. There are four subcases.

Subcase C1. $\frac{27}{106} \leq c \leq \frac{15}{53}$ so that $\frac{12}{53} \leq d \leq \frac{27}{106}$.

Captain Crook subdivides $a - \frac{27}{53} = \frac{26}{53} - b$ into $\frac{13}{53} - \frac{b}{2}$ and $\frac{13}{53} - \frac{b}{2}$. He will get three barrels with total volume $b + d + (\frac{13}{53} - \frac{b}{2}) \geq \frac{35}{53}$.

Subcase C2. $\frac{15}{53} \leq c \leq \frac{18}{53}$ so that $\frac{9}{53} \leq d \leq \frac{12}{53}$.

Captain Crook subdivides c into d and $c - d$. He will get three pieces with total volume $b + d + \min\{c - d, a - \frac{27}{53}\}$. In the former instance, it is at least $b + c \geq \frac{35}{53}$. In the latter instance, it is at least $d + 1 - \frac{27}{53} \geq \frac{35}{53}$.

Subcase C3. $\frac{18}{53} \leq c \leq \frac{24}{53}$ so that $\frac{3}{53} \leq d \leq \frac{9}{53}$.

Captain Crook subdivides c into $\frac{c}{2}$ and $\frac{c}{2}$. The crew gets two barrels with total volume at most $\frac{c}{2} + \max\{d, a - \frac{27}{53}\}$. In the former instance, it is at most $\frac{27}{53} - c \leq \frac{18}{35}$. In the latter instance, it is at most $\frac{12}{53} + \frac{6}{53} = \frac{18}{53}$.

Subcase C4. $\frac{24}{53} \leq c \leq \frac{27}{53}$ so that $0 \leq d \leq \frac{3}{53}$.

Captain Crook subdivides b into $\frac{b}{2}$ and $\frac{b}{2}$. He will get three barrels with total volume at least $c + \frac{n}{2} + \min\{d, a - \frac{27}{53}\}$. In the former instance, it is at least $\frac{27}{53} + \frac{b}{2} \geq \frac{37}{53}$. In the latter instance, it is at least $c + 1 - \frac{27}{53} - \frac{b}{2} \geq \frac{37}{53}$.

Case D. $\frac{1}{2} \leq a \leq \frac{27}{53}$ so that $\frac{26}{53} \leq b \leq \frac{1}{2}$.

Captain Crook passes. Whicever barrel the crew now subdivides, Captain Crook pour out from the larger of the two modified barrels an amount equal to $\frac{1}{3}$ of the barrel the crew has just subdivided. This will be the third largest barrel, and Captain Crook will get three barrels with total volume at least $b + \frac{a}{3} = 1 - \frac{2a}{3} \geq \frac{35}{53}$.

This completes the proof that Captain Crook can always get $\frac{35}{53}$ of the rum. We now prove that the crew can always get at least $\frac{18}{53}$ of the rum. We will need the following two preliminary results.

Lemma 1.

Suppose before the fourth and last move (by Captain Crook), the amount of rum in the first four barrels are $w \geq x \geq y \geq z$. If $x \leq 2y$, then the crew can get two barrels with total volume at least x.

Proof:

If Captain Crook subdivides either of the smallest two barrels, the second largest barrel will have volume x. There is nothing further to prove. Hence Captain Crook must subdivide either of the largest two barrels, into two barrels each with volume smaller than x. Because $x \leq 2y$, at least one of the modified barrels has volume less than y. If the original barrel with volume y is still the third largest, then the largest has volume at most w and the smallest has volume at most z. Hence Captain Crook gets three barrels with total volume at most $w + y + z$, so that the crew will get two pieces with total volume at least $1 - w - y - z = x$. On the other hand, if the barrel with volume y is now the second largest, then the volume of each of the two modified barrels lies between y and $x - y$. Thus the second smallest barrel has volume at least $x - y$, and the crew will get two barrels with total volume at least $y + (x - y) = x$.

Lemma 2.

Suppose before the fourth and last move (by Captain Crook), the amount of rum in the first four barrels are $w \geq x \geq y \geq z$. If $x \geq 2z$, then the crew can get two barrels with total volume at least $\min\{y + z, x + \frac{z}{2}\}$.

Proof:

If Captain Crook subdivides either of the smallest two barrels, the second smallest barrel will have volume at least $\frac{z}{2}$ while the second largest barrel will have volume x. Hence the crew will get two barrels with total volume at least $x + \frac{z}{2}$. If Captain Crook subdivides either of the largest two barrels into two barrels each with volume smaller than x, not both can have volume smaller than z since $x \geq 2z$. Hence the second smallest barrel has volume at least z while the second largest piece has volume at least y. Hence the crew will get two barrels with total volume at least $y + z$.

The crew's strategy begins with the crew subdivides 1 into $\frac{20}{53}$ and $\frac{33}{53}$. There are two cases.

Case A. Captain Crook subdivides $\frac{20}{33}$ into $a \geq b$.
There are three subcases.

Subcase A1. $\frac{18}{53} \leq a \leq \frac{20}{53}$ so that $0 \leq b \leq \frac{2}{53}$.
The crew subdivides $\frac{33}{53}$ into $\frac{18}{53}$ and $\frac{15}{53}$. In Lemma 1, let $w = a$, $x = \frac{18}{53}$, $y = \frac{15}{53}$ and $z = b$, with $x \leq 2y$. Hence the crew will get two barrels with total volume at least $\frac{18}{53}$.

Subcase A2. $\frac{17}{53} \leq a \leq \frac{18}{53}$ so that $\frac{2}{53} \leq b \leq \frac{3}{53}$.
The crew subdivides $\frac{33}{53}$ into $\frac{17}{53}$ and $\frac{16}{53}$. In Lemma 2, let $w = a$, $x = \frac{17}{53}$, $y = \frac{16}{53}$ and $z = b$, with $x \geq 2z$. Note that $y + z = \frac{16}{53} + b \geq \frac{18}{53}$ while $x + \frac{z}{2} = \frac{17}{53} + \frac{b}{2} \geq \frac{18}{53}$. Either way, the crew will get two barrels with total volume $\frac{18}{53}$.

Subcase A3. $\frac{10}{53} \le a \le \frac{17}{53}$ so that $\frac{3}{53} \le b \le \frac{10}{53}$.
The crew still subdivides $\frac{33}{53}$ into $\frac{17}{53}$ and $\frac{16}{53}$. The total volume of the smallest four barrel is $1 - \frac{17}{53} = \frac{36}{53}$. The crew is guaranteed to get at least half of that, which is $\frac{18}{53}$.

Case B. Captain Crook subdivides $\frac{33}{53}$ into $a \ge b$.
There are seven subcases.

Subcase B1. $\frac{27}{53} \le a \le \frac{33}{53}$, so that $0 \le b \le \frac{6}{53}$.
The crew subdivides a into $\frac{18}{53}$ and $a - \frac{18}{53}$. In Lemma 1, let $w = \frac{20}{53}$, $x = \frac{18}{53}$, $y = a - \frac{18}{53}$ and $z = b$, with $x \le 2y$. Hence the crew will get two barrels with total volume at least $\frac{18}{53}$.

Subcase B2. $\frac{51}{106} \le a \le \frac{27}{53}$, so that $\frac{6}{53} \le b \le \frac{15}{106}$.
The crew subdivides a into $\frac{15}{53}$ and $a - \frac{15}{53}$. In Lemma 2, let $w = \frac{20}{53}, x = \frac{15}{53}$, $y = a - \frac{15}{53}$ and $z = b$, with $x \ge 2z$. Now $y + z = a + b - \frac{15}{53} = \frac{33}{53} - \frac{15}{53} = \frac{18}{53}$ while $x + \frac{z}{2} = \frac{15}{53} + \frac{b}{2} \ge \frac{18}{53}$. Either way, the crew will get two barrels with total volume at least $\frac{18}{53}$.

Subcase B3. $\frac{25}{53} \le a \le \frac{51}{106}$, so that $\frac{15}{106} \le b \le \frac{8}{53}$.
The crew subdivides a into $a - b - \frac{3}{53}$ and $b + \frac{3}{53}$. If Captain Crook subdivides either $b + \frac{3}{53}$ or b, the second smallest barrel is at least $\frac{b}{2}$, and the crew will get two barrels with total volume at least $a - b - \frac{3}{53} + \frac{b}{2} = a + b - \frac{3}{53} - \frac{3b}{2} \ge \frac{18}{53}$. Suppose Captain Crook subdivides either $\frac{20}{53}$ or $a - b - \frac{3}{53}$. If both new barrels are less than b, then the crew will get two barrels with total volume at least $b + \frac{3}{53} + \frac{1}{2}(a - b - \frac{3}{53}) = \frac{18}{53}$. If at least one of the new barrels is greater than b, then the second largest barrel is at least $b + \frac{3}{53}$ so that the crew will get two barrels with total volume at least $b + \frac{3}{53} + b \ge \frac{18}{53}$.

Subcase B4. $\frac{23}{53} \le a \le \frac{25}{53}$, so that $\frac{8}{53} \le b \le \frac{10}{53}$.
The crew passes. If Captain Crook then subdivides b, the crew will get at least $\frac{20}{53}$. Hence he must subdivide a or $\frac{20}{53}$. After Captain Crook's final cut, if b is still the third largest, then he gets three barrels with total volume at most $a + b + 0 = \frac{33}{53}$, so that the crew will get two barrels with total volume at least $\frac{18}{53}$. If b becomes the second smallest, then the second largest is at least $\frac{10}{53}$, and the crew will get two barrels with total volume at least $\frac{10}{53} + b \ge \frac{18}{53}$.

Subcase B5. $\frac{20}{53} \le a \le \frac{23}{53}$, so that $\frac{10}{53} \le b \le \frac{13}{53}$.
The crew also passes. In Lemma 1, let $w = a$, $x = \frac{20}{53}$, $y = b$ and $z = 0$, with $x \le 2y$. Hence the crew will get two barrels with total volume at least $\frac{20}{53}$.

Subcase B6. $\frac{18}{53} \le a \le \frac{20}{53}$, so that $\frac{13}{53} \le b \le \frac{15}{53}$.
The crew still passes. In Lemma 1, let $w = \frac{20}{53}$, $x = a$, $y = b$ amd $z = 0$, with $x \le 2y$. Hence the crew will get two barrels with total volume at least $\frac{18}{53}$.

Subcase B7. $\frac{33}{106} \leq a \leq \frac{18}{53}$, so that $\frac{15}{53} \leq b \leq \frac{33}{106}$.
The crew subdivides $\frac{20}{53}$ into $\frac{14}{53}$ and $\frac{6}{53}$. In Lemma 2, let $w = a$, $x = b$, $y = \frac{14}{53}$ and $z = \frac{6}{53}$ with $x \geq 2z$. Now $y + z = \frac{20}{53}$ while $x + \frac{z}{2} = b + \frac{3}{53} \geq \frac{18}{53}$. Either way, the crew will get two barrels with total volume at least $\frac{18}{53}$.

This was the work [1] by Circle member Robert Barrington Leigh. Some editing and simplifications were made after his untimely death in 2006 by YunHao Fu, Zhichao Li and circle member **David Rhee**.

Exercises

1. Suppose the necklace is as shown in Figure 9.13. The values of the five jewels are 1, 2, 3, 4 and 5 in some order. What is the probability that the three jewels Captain Crook gets will be worth 8 or more?

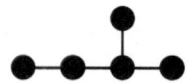

Figure 9.13

2. Each of m boxes contains some gold and some silver. The maximum amount of gold in any of the boxes is a kilograms and the maximum amount of silver in any of the boxes is b kilograms. Prove that the boxes can be divided into q groups such that the total number of boxes in one group differs from that in any other by at most 1, the total amount of gold in the boxes of one group differs from that of any other by at most a kilograms, and the total amount of silver in the boxes of one group differs from that of any other group by at most b kilograms.

3. If in the first stage of sharing rum, Captain Crook makes the first move instead of the crew, what is the maximum amount of rum he can get?

Bibliography

[1] Robert Barrington Leigh, YunHao Fu, Zhichao Li and David Rhee, A Surprising Result in Cake-Sharing, Crux Mathematicorum, **42** (2016) 252–257.

[2] Jean Bernard Landry Pellerin and David Rhee, How to be Fairly Greedy, submitted to *Ninth Gathering for Gardner Exchange Book*.

[3] Peter Winkler, *Mathematical Puzzles*, A K Peters Ltd., (2004) 1–2.

[4] Peter Winkler, *Mathematical Mind Benders*, A K Peters Ltd., (2007).

[5] Hsin-Po Wang, Sharing Eight Treasure Chests, *Eighth Gathering for Gardner Exchange Book*, Vol. 2 (2008) 181–190.

Chapter Ten: Puzzling Adventures

In this last chapter, we consider three problems from a delightful book [7]. For more problems which are important, instructive and interesting, see [8], [9], [10], [11] and [12] by the same author.

Section 1. Circuits Checking Circuits

A small airport has four ports A, B, C and D, but only one runway. Most of the time, at most one port will require the runway for either taking off or landing, so that no intervention from the control tower is necessary. It is desired to construct a circuit which would sound an alarm if and only if two or more ports want to use the runway at the same time.

The circuit consists of a number of gates each of which takes in a number of binary signals, that is, 0s and 1s, and combines them into a single one. There are two kinds of gates.

In Figure 10.1, an "OR" gate is illustrated on the left. Its output is 0 if all the inputs are 0s; otherwise, it is 1. An "AND" gate is illustrated on the right. Its output is 1 if all the inputs are 1s; otherwise, it is 0.

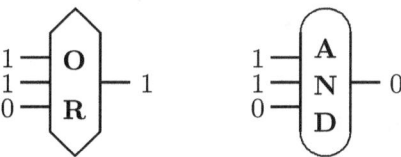

Figure 10.1

Each port will send out a binary signal: 0 if it does not require the runway, and 1 if it does. These inputs maybe fed into several gates of the circuit. An output from a gate may be fed into another gate further down the line. There is a final gate such that if its output is 0, this means no intervention is necessary, but if it is a 1, then an alarm will sound for the control tower to take charge.

A simplistic idea is to send each pair of inputs, namely (A,B), (A,C), (A,D), (B,C), (B,D) and (C,D) to a different AND gate, and send all six outputs into a final OR gate. Whenever two or more ports require the runway, at least one of the AND gates will output a 1, which will go through the OR gate as the final output. On the other hand, if at most one port requires the runway, then each AND gate outputs a 0 and so will the final OR gate.

This circuit has two drawbacks. First, it is not a good idea to have as many as six inputs converging on one gate. Second, if the number of ports increases linearly, the number of gates will increase quadratically.

© Springer International Publishing AG 2018
A. Liu, *S.M.A.R.T. Circle Projects*, Springer Texts
in Education, DOI 10.1007/978-3-319-56811-9_10

Figure 10.2 shows a much better design, even though it also uses seven gates.

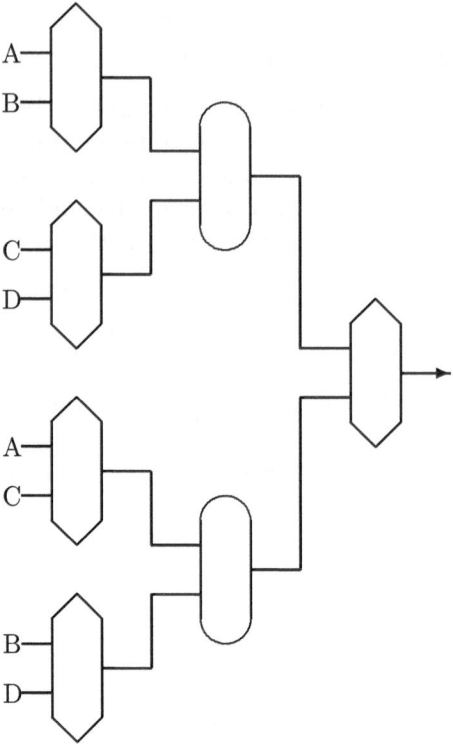

Figure 10.2

If two or more gates require the runway and they include the pair (A,C), (A,D), (B,C) or (B,D), a 1 will be fed into the final OR gate from the upper branch. If they include (A,B), (A,D), (C,B) and (C,D), a 1 will be fed into the final OR gate from the lower branch.

Note that the cases (A,D) and (B,C) are checked twice. Using this fact, it is possible to reduce the number of gates used to six. However, to reduce it further requires a new idea. This time, we feed the triples (A,B,C), (A,B,D), (A,C,D) and (B,C,D) to a different OR gate, and send all four outputs into a final AND gate.

We can avoid having as many as four inputs converging on one gate by modifying this circuit into the one shown in Figure 10.3. If only A or only B requires the runway, a 0 will be fed into the final AND gate from the upper branch. If only C or only D requires the runway, a 0 will be fed into the final AND gate from the lower branch. If two or more gates require the runway, a 1 will be fed into the final AND gate from the upper branch as well as from the lower branch.

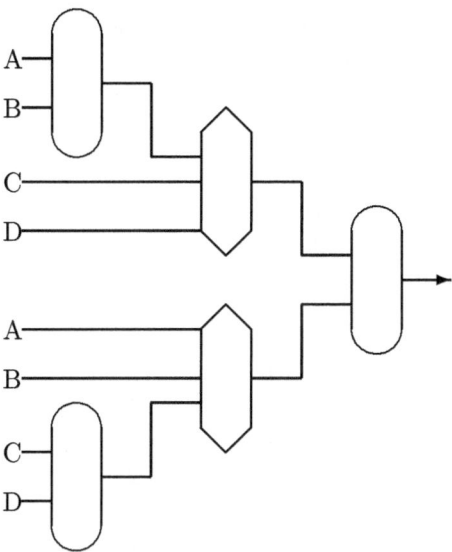

Figure 10.3

We now turn to a more difficult problem. A medium-sized airport has nine ports A, B, C, D, E, F, G, H and I, and two runways. Most of the time, at most two ports will require the runways for either taking off or landing, so that no intervention from the control tower is necessary. It is desired to construct a circuit which would sound an alarm if and only if three or more ports want to use the runway at the same time.

This problem turns out to be isomorphic to a beautiful problem proposed by Hungary for the 1988 International Mathematical Olympiad in Australia. It was put on the short-list (see [2]) by the Problem Committee, but did not make it to the contest paper.

In a multiple-choice test there were 4 questions and 3 possible answers for each question. A group of students was tested and it turned out that for any 3 of them, there was a question which the three students answered differently. What is the maximal possible number of students tested?

If three students answer a question differently, we say that they are distinguishable by that question. Denote by $f(n)$ the maximal possible number of students in such a test with n questions. We have $f(1) = 3$ since the three students can give different answers. If we have four or more students, two of them must give the same answer by the Pigeonhole Principle. Adding any third student will form a trio not distinguishable by the lone question.

For $n \geq 1$, we have $f(n + 1) \leq \frac{3}{2}f(n)$. Suppose $f(n + 1) = m$. Consider the $(n + 1)$-st question. At least $\frac{2m}{3}$ of the students did not give the less popular answer for that question. Then these students must be distinguishable by at least one of the first n questions. Hence $f(n) \geq \frac{2m}{3}$ or $f(n + 1) \leq \frac{3}{2}f(n)$.

Now $f(2) \leq \frac{3}{2}f(1) \leq 4$, $f(3) \leq \frac{3}{2}f(2) \leq 6$ and $f(r) \leq \frac{3}{2}f(3) \leq 9$. In the chart below, we show the responses to the four questions #1, #2, #3 and #4 by nine students A, B, C, D, E, F, G, H and I.

Students	#1	#2	#3	#4
A	0	0	0	0
B	0	1	1	1
C	0	2	2	2
D	1	0	2	1
E	1	1	0	2
F	1	2	1	0
G	2	0	1	2
H	2	1	2	0
I	2	2	0	1

We claim that every three students are distinguishable by at least one question. Arrange their seats in a 3×3 array as follows.

$$
\begin{array}{ccc}
A & B & C \\
D & E & F \\
G & H & I
\end{array}
$$

If three students sit in different rows, they are distinguishable by #1. If they sit in different columns, they are distinguishable by #2. If they sit in diffenent down diagonals (A-E-I, B-F-G and C-D-H), they are distinguishable by #3. Finally, if they sit in different up-diagonals (A-F-H, B-D-I and C-E-G), they are distinguishable by #4. Since there are only three pairs of students among a trio, they will be distinguishable by at least one of the four questions.

So $f(4) = 9$. It follows that the earlier upper bounds for $f(2)$ and $f(3)$ are also exact. In other words, we have $f(2) = 4$ and $f(3) = 6$. It can be proved that $f(5) = 10$, a surprisingly low value which indicates that the general problem of determining $f(n)$ is very difficult.

We now return to the circuit problem. We will feed the nine inputs into three OR gates, with three going into each. Then we feed the three outputs into an AND gate. This we do into four different ways, and feed the four outputs from the AND gates into a final OR gate. This circuit is shown in Figure 10.4 which uses 17 gates, an amazingly low number.

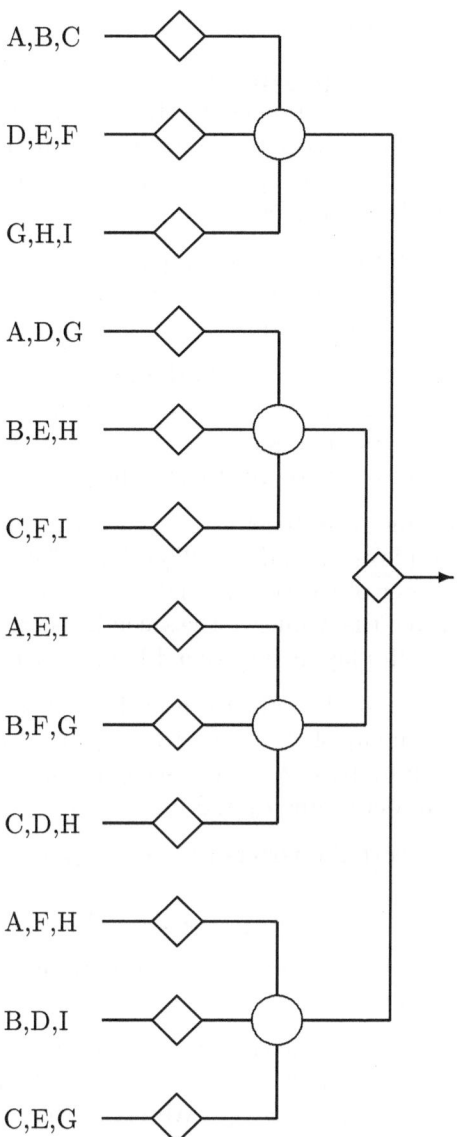

A,B,C

D,E,F

G,H,I

A,D,G

B,E,H

C,F,I

A,E,I

B,F,G

C,D,H

A,F,H

B,D,I

C,E,G

Figure 10.4

This is based on the work of Circle member Graham Denham. See [3] and [4].

Section 2. The Coach's Dilemma.

A tennis coach has eight players, all of different strengths. The coach wishes to rank them from the strongest to the weakest. It takes exactly one hour of playing against each other to determine which of two players is the stronger. As many as four pairs can be playing during the same hour. The task must be completed within 6 hours.

A standard approach for solving this type of problem is divide and conquer. We first divide the eight players into two groups of four, and then subdivide each group into two pairs.

The players in a pair go against each other, resulting in four sorted pairs in 1 hour. Next, we perform simultaneously two mergers of two sorted pairs (a_1, a_2) and (b_1, b_2) into a sorted quartet (c_1, c_2, c_3, c_4). This can be done in at most 2 hours. We present two different methods.

The first method, due to Batcher [2], is called the odd-even merge-sort. In the first hour, a_1 plays b_1 and a_2 plays b_2. The winner of the first match is c_1, and the loser of the second match is c_4. If the loser of the first match has already played the winner of the second, the ranking is completed. Otherwise, these two will play in the second hour to determine c_2 and c_3.

The second method may be called the upside-down merge-sort. In the first hour, a_1 plays b_2 and b_1 plays a_2. If either a_2 or b_2 wins, the ranking is completed. If not, then a_1 plays b_1 in the second hour to determine c_1 and c_2, while a_2 plays b_2 to determine c_3 and c_4.

We now merge two sorted quartets (c_1, c_2, c_3, c_4) and (d_1, d_2, d_3, d_4) into a sorted octet

$$(e_1, e_2, e_3, e_4, e_5, e_6, e_7, e_8).$$

This can be done in at most 3 hours, so that the overall task can be accomplished in at most 6 hours.

Using Batcher's method, it takes at most 2 hours to merge (c_1, c_3) with (d_1, d_3) into (f_1, f_2, f_3, f_4) as well as (c_2, c_4) and (d_1, d_3) into (g_1, g_2, g_3, g_4). Clearly, we have $f_1 = e_1$ and $g_4 = e_8$. We claim that $\{f_2, g_1\} = \{e_2, e_3\}$, $\{f_3, g_2\} = \{e_4, e_5\}$ and $\{f_4, g_3\} = \{e_6, e_7\}$, so that at most 1 more hour is needed for completing the ranking.

By symmetry, we may assume that $f_1 = c_1$. Then $f_2 = d_1$ or c_3 while $g_1 = d_2$ or c_2. Now only c_1 and possibly c_2 can be ahead of f_2. Similarly, only c_1 and possibly d_1 can be ahead of g_1. Hence $\{f_2, g_1\} = \{e_2, e_3\}$. By symmetry again, $\{f_4, g_3\} = \{e_6, e_7\}$, so that we have $\{f_3, g_2\} = \{e_4, e_5\}$ as desired.

Using our method, we have c_1 play d_4, c_2 play d_3, c_3 play d_2 and c_4 play d_1 in the first hour. If c_4 or d_4 wins, the ranking is completed. So we may as well assume that c_1 and d_1 win.

If c_2 and d_2 also win, we can complete the ranking by merging (c_1, c_2) with (d_1, d_2) to yield (e_1, e_2, e_3, e_4), as well as (c_3, c_4) with (d_3, d_4) to yield (e_5, e_6, e_7, e_8). This takes at most 2 more hours. Suppose c_3 beats d_2. Then $\{c_1, c_2, c_3, d_1\} = \{e_1, e_2, e_3, e_4\}$ and $\{c_4, d_2, d_3, d_4\} = \{e_5, e_6, e_7, e_8\}$. In the second hour, d_1 plays c_2 while c_4 plays d_3. In the third hour, d_1 plays c_1 following a victory or c_3 following a defeat, while c_4 plays d_2 following a victory or d_4 following a defeat.

In the general case using Batcher's method, let there be 2^n as and 2^n bs. Let the two sequences be sorted separately. Let the odd-indexed terms be sorted into cs and the even-indexed terms be sorted into ds. We claim that in the final sorting into es, $\{e_{2k}, e_{2k+1}\} = \{c_{k+1}, d_k\}$ for $1 \le k \le 2^{n+1} - 1$. This is proved in many standard textbooks on parallel algorithms. See for example [1]. Here we give our own proof via two auxiliary results.

Lemma 1.
For $1 \le k \le 2^n$, c_k is ahead of d_k.

Proof:
By symmetry, we may assume that $d_k = a_{2m}$ for some m. Among the ds, there are $m - 1$ other as ahead of d_k, namely, $a_2, a_4, \ldots, a_{2m-2}$. Hence there are $k - m$ bs ahead of d_k, namely, $b_2, b_4, \ldots, b_{2k-2m}$. Among the cs, $a_1, a_3, \ldots, a_{2m-1}, b_1, b_3, \ldots, b_{2k-2m-1}$ are ahead of d_k. These consist of m as and $k - m$ bs, so that $m + (k - m) = k$ cs are ahead of d_k. Since c_k is in the kth place among the cs, c_k is ahead of d_k.

Lemma 2.
For $2 \le k \le 2^n - 1$, d_{k-1} is ahead of c_{k+1}.

Proof:
By symmetry, we may assume that $c_{k+1} = a_{2m-1}$ for some m. Among the cs, there are $m - 1$ other as ahead of c_{k+1}, namely, $a_1, a_3, \ldots, a_{2m-3}$. Hence there are $k - m + 1$ bs ahead of c_{k+1}, namely, $b_1, b_3, \ldots, b_{2k-2m+1}$. Among the ds, $a_2, a_4, \ldots, a_{2m-2}, b_2, b_4, \ldots, b_{2k-2m}$ are ahead of c_{k+1}. These consist of $m - 1$ as and $k - m$ bs, so that $(m - 1) + (k - m) = k - 1$ ds are ahead of c_{k+1}. Since d_{k-1} is in the $(k-1)$st place among the ds, d_{k-1} is ahead of c_{k+1}.

We give an illustration using $k = 5$. Lemma 1 states that c_5 is ahead of d_5. Note that d_5 is either an a or a b. By symmetry, we may assume d_5 is an a, and let us say $d_5 = a_6$. Then it is the third a in the even group, so that it is behind b_2 and b_4. It follows that d_5 is behind a_1, a_3, a_5, b_1 and b_3 in the odd group, and one of these is c_5.

Lemma 2 states that d_4 is ahead of c_6. Note that c_6 is either an a or a b. By symmetry, we may assume c_6 is an a, and let us say $c_6 = a_3$. Then it is the second a in the odd group, so that it is behind b_1, b_3, b_5 and b_7. It follows that c_6 is behind a_2, b_2, b_4 and b_6 in the even group, and one of these is d_4.

The two lemmas together show that after the cs and ds have been sorted, only 1 extra hour is needed to complete the ranking.

We now prove that our upside-down merge-sort also works in general.

Theorem.
If $k + t = 2^n$, $(a_1, a_2, \ldots, a_k)$ can be merged with $(b_1, b_2, \ldots, b_t)$ in at most n hours.

Proof:
We use induction on n. The case $n = 1$ is trivial. Suppose the result holds for some $n \geq 1$. Consider the next case where $k + t = 2^{n+1}$. We may assume that $1 \leq k \leq t$. For $1 \leq i \leq k$, let a_i play b_{m-i+1} in the first hour, where $m = 2^n$. We consider three cases.

Case 1. All the as win.
In particular, a_k beats b_{m-k+1}. In order to complete the ranking, we only need to merge $(a_1, a_2, \ldots, a_k)$ with $(b_1, b_2, \ldots, b_{m-k})$. Since $k + (m-k) = 2^n$, at most n more hours are needed by the induction hypothesis.

Case 2. All the as lose.
In particular, b_m beats a_1. In order to complete the ranking, we only need to merge $(a_1, a_2, \ldots, a_k)$ with $(b_{m+1}, b_{m+2}, \ldots, b_t)$. Since $k + (t - m) = 2^n$, at most n more hours are needed by the induction hypothesis.

Case 3. Not all as win and not all as lose.
Let $i < k$ be the largest index such that a_i wins. In order to complete the ranking, we only need to merge $(a_1, a_2, \ldots, a_i)$ with $(b_1, b_2, \ldots, b_{m-i})$, and at the same time $(a_{i+1}, a_{i+2}, \ldots, a_k)$ with $(b_{m-i+1}, b_{m-i+2}, \ldots, b_t)$. Since $i + (m - i) = 2^n = (t - (m - i)) + (k - i)$, at most n more hours are needed by the induction hypothesis.
In all cases, $n + 1$ hours are sufficient, completing the inductive proof of our Theorem.

This problem also appeared in the International Mathematics Tournament of Towns [13] with more players. Circle member Calvin Li [6] discovered the upside-down merge-sort while writing the contest.

Section 3. The Campers' Problem

Suppose that you are a camp counselor. You and your eight campers are lost in the woods unable to find a path. Finally, you come to a four-way intersection of paths: N, E, W or S. You know your campsite is only 20 minutes away from there, but you don't know which path to take. You have an hour more of daylight, after which traveling is very dangerous. So you cannot travel with all eight campers down one route at a time. It would take too long.

Indeed, you must send small groups 20 minutes down each path and have them rendezvous at the intersection in 40 minutes. You will then decide which route to take. You may also participate in the search in the first 40 minutes.

The problem is that two of the campers in your group sometimes lie. You do not know which ones they are. How do you divide up your group into search parties? At rendezvous time, how do you decide which way to go? You must be right no matter how the occasional liars—whoever they may be—are distributed among the groups and no matter whether they lie or not.

Certainly, you can check out one path, say N. Perhaps you send four campers down each of E and W. The trouble is that one group may agree unanimously that the camp is not there, while the other group returns a split decision of two against two. You know for sure that the liars are one pair in that group, but you do not know which pair.

The key observation is that a unanimous decision among three campers can be accepted as truth, so that sending a fourth person down a path is wasteful. So we modify our strategy and send three campers down each of E and W, and the remaining two down S.

There seem to be a lot of cases which we have to consider, according to who says "Yes" and who says "No". However, the nature of the answer is immaterial. The important thing is whether we have a unanimous decision or a split decision. You can make the call based on the chart below.

E Group	W Group	S Group || Believe
3:0	3:0	2:0 or 1:1	E and W Groups
3:0	2:1	1:1	E and W Groups
3:0	2:1	2:0	E and S Groups
2:1	2:1	2:0	E and W Groups.

When a split group is believed, it is understood that the majority is believed.

So far, this sounds more like an exercise in logic than a problem in mathematics. We will address this issue in three ways.

First, we want to determine the minimum value of the total number of campers needed if two of them are liars. Can the original problem be solved if the total number of campers is seven?

We assume as before that you go down N and do not find the camp. With seven campers distributed among the remaining three paths, the Pigeonhole Principle tells us that at least three campers must go down one of them. This means that two of them, say E and W, must be visited by only four campers.

Suppose the camp is not found down S either. If the distribution is four for E and zero for W, then the group of four can come back split, two saying yes and two saying no. You will not know whether E ss right or W ss right. If the distribution is three for E and one for W, then any disagreement in E also leaves open both possibilities. If the distribution is two and two, then if both come back with the same answer, you again have no way to decide.

We can give a simpler argument, using symmetry, because both liars may be among the four campers visiting E and W, and here the truth-teller do not have a majority. Whatever the two truth-tellers can convince you, the two liars can convince you the opposite. Thus you cannot tell between them.

The second way to highlight the mathematical flavor of the problem is to see that it fits into a pattern. If there are no liars, clearly two campers would be necessary and sufficient.

Suppose there is one liar. The argument above involving the Pigeonhole Principle and symmetry tells us that four campers in all will not suffice. With five campers, we can send two down each of E and W, and the remaining one down S. If both groups of two are unanimous, believe them. If one of them is split, believe the other two groups. They cannot both be split since there is only one liar.

In general, suppose we have n liars. We know that we need $3n + 2$ campers in all. We plan to send $n + 1$ of them down each of E and W, and the remaining n down S. We know what is a unanimous decision, but there are way too many possible split decisions. How we can handle them?

When the group has been reassembled, let M_s be the number of campers in the majority and m_s be the number of campers in the minority among those down S. Note that we may have $M_s = m_s$. Let M_e, m_e, M_w and m_w be defined in an analogous manner.

If $\sum_s = M_s + m_e + m_w > n$, you can believe the M_s campers. This is because the E group contains at least m_e liars and the W group contains at least m_w liars. If you cannot believe the M_s campers, then they must all be liars. However, there are only n liars, and we cannot have $\sum_s > n$.

Similarly, if $\sum_e = M_e + m_w + m_s > n$, you can believe the M_e campers. If $\sum_w = M_w + m_s + m_e > n$, you can believe the M_w campers. If you have doubts about at most one of M_s, M_e and M_w, you can deduce where the campsite is. Suppose you have doubts about M_e and M_w. This means that $n \geq \sum_e$ and $n \geq \sum_w$. Addition yields

$$2n \geq (M_e + m_e) + (M_w + m_w) + 2m_s = 2n + 2 + 2m_s,$$

which is a contradiction. Doubting M_s and M_e, or doubting M_s and M_w, leads to a similar contradiction.

Let us interpret the above argument with the original case $n = 2$.

M_e	m_e	M_w	m_w	M_s	m_s	$\sum_e$	$\sum_w$	$\sum_s$	Believe
3	0	3	0	2	0	**3**	**3**	2	M_e and M_w
3	0	3	0	1	1	**4**	**4**	1	M_e and M_w
3	0	2	1	1	1	**5**	**3**	2	M_e and M_w
3	0	2	1	2	0	**4**	2	**3**	M_e and M_s
2	1	2	1	2	0	**3**	**3**	**4**	all three

Finally, we show that this problem is closely related to Error-Correcting Codes, the topic of the first chapter of this book.

Suppose we have a transmitter which sends out 8 binary digits at a time, with at most two digit-reversal per transmission. We use the first and fourth digits as the message, which must be one of 00, 01, 10 or 11. We copy the first digit two more times as the second and third digits and copy the second digit two more times as the fifth and sixth digits. The last two digits are 0s if the first two digits are the same, and are 1s if the first two digits are not the same. The chart below shows the encoded messages.

0	0	0	**0**	0	0	0	0
0	0	0	**1**	1	1	0	0
1	1	1	**0**	0	0	1	1
1	1	1	**1**	1	1	1	1

If up to two errors occur, we may correct them following the same reasoning in the problem with the eight campers.

Exercises

1. Modify the circuit in Figure 10.2 into a circuit using only six gates.

2. Rank the eight players in 17 hours if only one pair can play in the same hour.

3. Solve the problem using four campers, two of whom sometimes lie. You have time for two exploratory trips and one final walk to the site.

Bibliography

[1] S. Akl, *The Design and Analysis of Parallel Algorithms*, Prentice-Hall (1989) 60–64 and 87–89.

[2] K. Batcher, Sorting Networks and their Applications, in *Proceedings of the AFIPS 32nd Spring Joint Computing Conference*, (1968) 307–314.

[3] Graham Denham and Andy Liu, Circuits Checking Circuits, in *Mathematics in Education*, ed. T. Rassias, University of La Verne (1992) 145–150.

[4] Graham Denham and Andy Liu, Competitions, Matrices, Geometry and Circuits, Austral. Math. Gaz. **24** (1997) 109–113.

[5] W. P. Galvin, D. C. Hunt and P. J. O'Halloran, *An Olympiad Down Under*, Australian Mathematics Foundation Ltd., Canberra (1988) 65 and 99.

[6] Calvin Li and Andy Liu, The Coach's Dilemma, Mathematics and Informatics Quarterly **2** (1992) 155–157.

[7] Dennis Shasha, *The Puzzling Adventures of Dr. Ecco*, Dover Publications Inc., Mineola (1998) 23–26, 49–53, 59–62, 141–145, 151–153 and 155–156.

[8] Dennis Shasha, *Dr. Ecco, Mathematical Detective*, Dover Publications Inc., Mineola (2004).

[9] Dennis Shasha, *Dr. Ecco's Cyberpuzzles*, W. W. Norton (2002).

[10] Dennis Shasha, *Puzzling Adventures*, W. W. Norton (2005).

[11] Dennis Shasha, *The Puzzler's Elusion*, Thunder's Mouth Press (2006).

[12] Dennis Shasha, *Puzzles for Programmers and Pros*, Wiley (2007).

[13] Peter Taylor, *International Mathematics Tournament of the Towns: Book 3: 1989–1993*, Australian Mathematics Trust, Canberra (1994) 70, 76, 89–90 and 105.

Appendix A: Additional Problems

Problems

1. In the convex quadrilateral $ABCD$, AC meets BD at E. If $AB = CD$ and $AE = CE$, is $ABCD$ necessarily a parallelogram?

2. X is a point inside a square $ABCD$ such that $\angle XCD = \angle XDC = 15°$. Give a direct proof that triangle ABX is equilateral.

3. Prove that $\dfrac{1}{\sin\frac{\pi}{7}} = \dfrac{1}{\sin\frac{2\pi}{7}} + \dfrac{1}{\sin\frac{3\pi}{7}}$.

4. Let ABC be a triangle. L, M and N are collinear points lying on the lines BC, CA and AB respectively. Construct the point P such that the lines AP, BP and CP intersect the lines BC, CA and AB at D, E and F, respectively, and that the lines EF, FD and DE pass through L, M and N, respectively.

5. Find the angles of all triangles which can be dissected into two isosceles triangles.

6. Dissect a square of suitable size into 6 similar rectangles so that there are 1 large rectangle, 3 medium rectangles and 2 small rectangles, each with integral dimensions.

7. There is a 3×10 hole of depth $\frac{1}{2}$ on a wall. In how many ways can it be filled by 15 rectangular bricks each of thickness $\frac{1}{2}$ and cross-section 1×2 or 2×1?

8. A V-tromino is a 2×2 square with one of the 4 1×1 square missing. Prove that if any of the 49 1×1 square is removed from a 7×7 board, the remaining part can be covered by 16 copies of the V-tromino.

9. An L-tetromino is a 2×3 rectangle missing one corner square and an adjacent non-corner square. Determine the minimum number n of colors for which there exists an n-color infinite infinite chessboard such that wherever an L-tetromino is placed, the four squares it covers all have different colors.

10. Consider a block of three stamps A, B and C in a row. It has six connected subblocks, namely, A, B, C, AB, BC and ABC. If the values of A, B and C are 1, 3 and 2 respectively, then the values of the six blocks are six consecutive integers, starting with 1. Such a block of stamps is said to be perfect. Find a perfect blocks consisting of five stamps, not necessarily in a row. Two adjacent stamps must meet along an entire edge.

© Springer International Publishing AG 2018
A. Liu, *S.M.A.R.T. Circle Projects*, Springer Texts
in Education, DOI 10.1007/978-3-319-56811-9

11. A G4G7 number is defined to be a positive integer with at least 4 digits such that each digit is at least 7, that is, each is 7, 8 or 9. Find infinitely many pairs of G4G7 numbers whose products are also G4G7 numbers.

12. A and B are 20 kilometers apart, and are moving towards each other, both at 10 kilometers per hour. At the same time, a bee starts off from A and flies towards B. When it reaches B, it immediately turns around and flies back towards A. It goes back and forth between A and B until they meet. When flying from A towards B, the bee's speed is 18 kilometers per hour. When flying from B to A, it is 12 kilometer per hour. How far has the bee flown when A and B meet?

13. A girls' school had 18 students in the graduating class. They invited a number of students from a boys' school nearby to their graduation dance. Each of the 18 girls danced with at most 3 boys. Each boy danced with at least 3 girls. What was the minimum number of girls such that each boy had danced with at least one of them?

14. Seated in a circle are 8 wizards. A positive integer not exceeding 100 is pasted onto the forehead of each. The numbers need not be distinct. A wizard can see the numbers of the other 7, but not his own. After ten minutes, a bell rings. Simultaneously, each wizard puts up either his left hand or his right hand. After another ten minutes, the bell rings again. Simultaneously, each wizard declares the number on his forehead. Find a strategy on which the wizards can agree beforehand, which can guarantee that each of them will make the correct declaration?

15. A tournament is a complete directed graph, meaning that between any two vertices, there is exactly one arc pointing from one of them to the other. Three vertices X, Y and Z are said to form a 3-cycle if the arcs go from X to Y to Z and back to X, or X to Z to Y and back to X. In a tournament with n players, let the number of arcs going out of the i-th vertex be w_i for $1 \le i \le n$. Prove that the number of 3-cycles is given by

$$\frac{n(n-1)(2n-1)}{12} - \frac{1}{2} \sum_{i=1}^{n} w_i^2.$$

16. How many convex polyhedra are there with five vertices?

Remark:
The problems in the Appendix are based on the work of Circle members, whose names are in **boldface**, and members of Chiu Chang Mathematical Circle, whose names are in *italic*.

Bibliography

1. **Brandan Capel** and **Alan Tsay**, Three Solutions to a Geometry Problem, Π in the Sky, September (2002) 29.

2. **Frank Chen, Anton Cherney** and **Kenneth Nearey**, Dissecting Rectangular Strips into Dominoes, *Mathematics for Gifted Students II*, Alberta Teachers' Association (1996) 27-29.

3. **Jack Chen**, Bargain Hunting in Geometown, Edu Math **34** (2012) 57–71.

4. *Sheng-Yuan Chen*, Generalization of von Neumann's Bee Problem, Edu Math **14** (2002) 45–49 (in Chinese).

5. *Daniel Chiu, Thomas Yao* and *Hans Yu*, Covering square boards with V-trominoes, Pi in the Sky **17** (2013) 28–30.

6. **Byung Kyu Chun**, Andy Liu and **Daniel van Vliet**, Dissecting Squares into Similar Rectangles, Crux Math. **22** (1996) 241-248.

7. **Steven Laffin**, An Imaginative Postal Service, in *Mathematics for Gifted Students II*, Alberta Teachers' Association (1996) 23-25.

8. *Jerry Lo* and **David Rhee**, Non-transitivity in Tournaments, Crux Mathematicorum with Mayhem, **32** (2006) 298–302.

9. *Jerry Lo* and **David Rhee**, Trilinear Poles and Polars, Integral **9** (2006) 58–68.

10. **Richard Ng**, Polyhedra with six vertices, Π *in the Sky*, September (2002), 26–27.

11. **David Rhee** and Andy Liu, Seven on Seven, *Seventh Gathering for Gardner Exchange Book* Volume 1 (2006) 297–300.

12. **David Rhee** and Andy Liu, Four on Seven, *Seventh Gathering for Gardner Exchange Book* Volume 2 (2006) 169–170.

13. **Daniel Robbins, Sudhakar Sivapalan** and **Matthew Wong**, How to Flip without Flipping, *Mathematics for Gifted Students II*, Alberta Teachers' Association (1996) 33-34.

14. **Mariya Sardarli**, The Graduation Dance, submitted to *Ninth Gathering for Gardner Exchange Book*.

15. *Hans Yu*, Polyominoes on a Multicolored Infinite Grid, Mathematics Competitions **27** (2014) #1 58–66.

16. **Jonathan Zung**, A Magic Trick with Eight Wizards, *Eighth Gathering for Gardner Exchange Book* Volume 1 (2008) 143–145.

Solution to Problems

1. In the convex quadrilateral $ABCD$, AC meets BD at E. If $AB = CD$ and $AE = CE$, is $ABCD$ necessarily a parallelogram?

   Solution:
   The answer is "No". We construct a counter-example as follows. Take any triangle with two equal sides and any point on its third side but not its midpoint. Join this point to the opposite vertex and cut along this line segment to make two triangles, labeled as shown in Figure A.1. Note that $\angle CFD = \angle AEB$.

   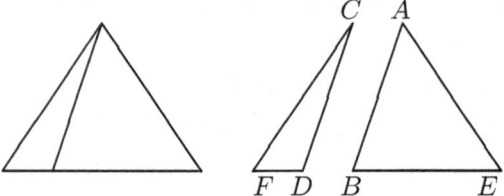

   Figure A.1

   Identify the point F with E and make it the point of intersection of AC and BD as shown in Figure A.2. Then we have $AB = CD$ and $AE = CE$, but $ABCD$ is not a parallelogram.

   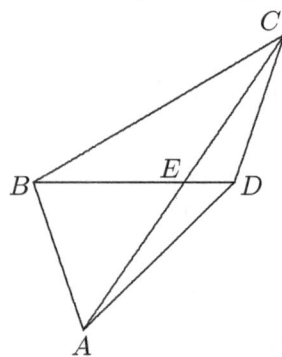

   Figure A.2

2. X is a point inside a square $ABCD$ such that $\angle XCD = \angle XDC = 15°$. Give a direct proof that triangle ABX is equilateral.

Solution:
Let Y be the point inside triangle DAX such that DXY is equilateral, as shown in Figure A.3. Then $\angle XDY = 60°$. It follows that we have $\angle ADY = 15° = \angle CDX$. Since $DY = DX$ and $DA = DC$, triangles DAY and DCX are congruent, so that

$$\angle AYD = \angle CXD = 180° - \angle XCD - \angle XDC = 150°.$$

It follows that

$$\angle AYX = 360° - \angle AYX - \angle DYX = 150°.$$

Moreover, $AY = CX = DX = XY$, so that

$$\angle AXY = \angle XAY = \frac{1}{2}(180° - \angle AYX) = 15°.$$

Hence $\angle ADX = \angle AXD$, so that $AX = AD$. By symmetry, we have $AX = BX$. Hence ABX is indeed an equilateral triangle.

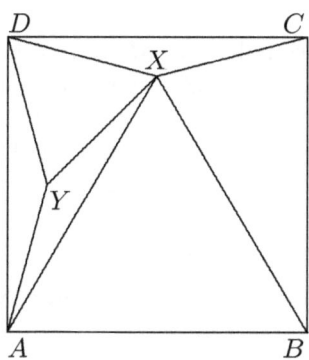

Figure A.3

3. Prove that $\dfrac{1}{\sin \frac{\pi}{7}} = \dfrac{1}{\sin \frac{2\pi}{7}} + \dfrac{1}{\sin \frac{3\pi}{7}}$.

Solution:
Figure A.4 shows a convex quadrilateral whose vertices are four of the seven vertices of a regular heptagon inscribed in a circle of radius R. By the Law of Sines, its sides and diagonals have lengths $a = 2R \sin \dfrac{\pi}{7}$, $b = 2R \sin \dfrac{2\pi}{7}$ and $c = 2R \sin \dfrac{3\pi}{7}$. By Ptolemy's Theorem, $bc = ca + cb$, and this is equivalent to $\frac{1}{a} = \frac{1}{b} + \frac{1}{c}$. Multiplying by $2R$ yields the desired result.

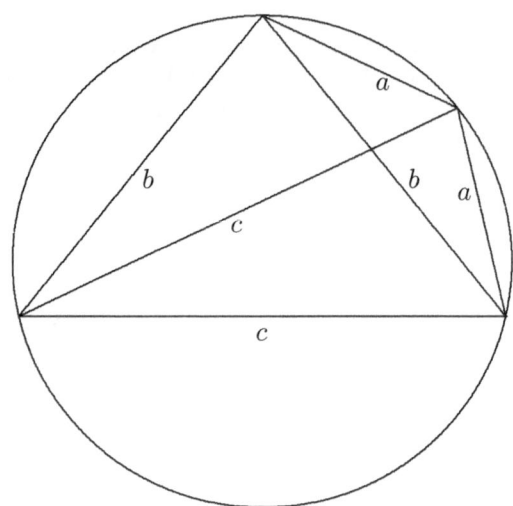

Figure A.4

4. Let ABC be a triangle. L, M and N are collinear points lying on the lines BC, CA and AB respectively. Construct the point P such that the lines AP, BP and CP intersect the lines BC, CA and AB at D, E and F, respectively, and that the lines EF, FD and DE pass through L, M and N, respectively.

Solution:

Through L, draw any line intersecting CA at H and AB at K. Let Q be the point of intersection of BH and CK and D be that of AQ and BC, as shown in Figure A.5.

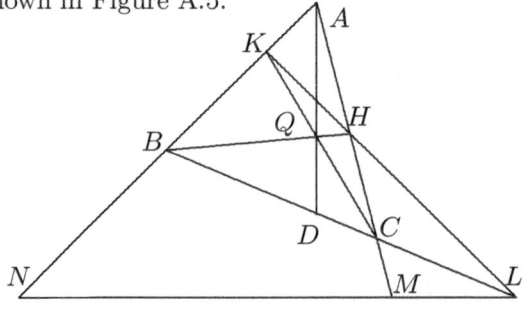

Figure A.5

Applying Ceva's Theorem to triangle ABC with respect to the point Q, we have

$$\frac{CD}{DB} \cdot \frac{BK}{KA} \cdot \frac{AH}{HC} = 1. \tag{10}$$

Applying Menelaus' Theorem to triangle ABC with respect to the lines LMN and HKL, we have

$$\frac{CL}{LB} \cdot \frac{BN}{NA} \cdot \frac{AM}{MC} = -1, \tag{11}$$

$$\frac{BL}{LC} \cdot \frac{CH}{HA} \cdot \frac{AK}{KB} = -1. \tag{12}$$

Let F be the point of intersection of DM and AB, P be that of CF and AD, and E be the point of intersection of DN and AM, as shown in Figure A.6. We claim that E lies on FL and P lies on BE.

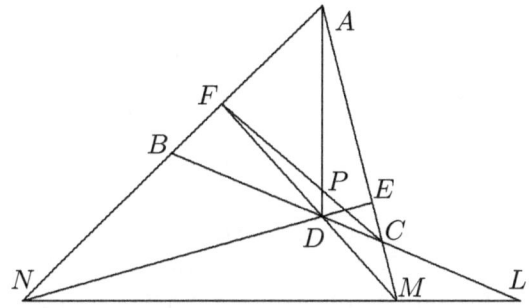

Figure A.6

Applying Menelaus' Theorem to triangle ABC with respect to the lines DFM and DEN, we have

$$\frac{BD}{DC} \cdot \frac{CM}{MA} \cdot \frac{AF}{FB} = -1, \tag{13}$$

$$\frac{BD}{DC} \cdot \frac{CE}{EA} \cdot \frac{AN}{NB} = -1. \tag{14}$$

Applying Menelaus' Theorem to triangle BAD with respect to the line PCF, we have

$$\frac{DC}{CB} \cdot \frac{BF}{FA} \cdot \frac{AP}{PD} = -1. \tag{15}$$

Multiplying (1), (1), (2), (3), (3), (4) and (5), we have

$$\frac{BL}{LC} \cdot \frac{CE}{EA} \cdot \frac{AF}{FB} = -1.$$

Applying the converse of Menelaus' Theorem to triangle ABC, E lies on LF. Multiplying (1), (2), (3), (4), (5) and (6), we have

$$\frac{CB}{BD} \cdot \frac{DP}{PA} \cdot \frac{AE}{EC} = -1.$$

Applying the converse of Menelaus' Theorem to triangle CAD, P lies on BE.

5. Find the angles of all triangles which can be dissected into two isosceles triangles.

Solution:
Clearly, such a triangle can only be cut into two triangles by drawing a line from a vertex to the opposite side, as illustrated in Figure A.7.

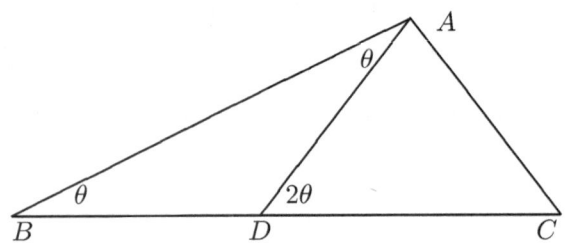

Figure A.7

Note that at least one of $\angle ADB$ and $\angle ADC$ is non-acute. We may assume that $\angle ADB \geq 90°$. In order for BAD to be an isosceles triangle, we must have $\angle BAD = \angle ABD$. Denote their common value by θ. By the Exterior Angle Theorem, $\angle ADC = 2\theta$. There are three ways in which CAD may become an isosceles triangle.

Case 1. $\angle ACD = \angle ADC = 2\theta$.
Then $\angle CAD = 180° - 4\theta > 0°$. This class consists of all triangles in which two of the angles are in the ratio 1:2, where the smaller angle θ satisfies $0° < \theta < 45°$.

Case 2. $\angle CAD = \angle ADC = 2\theta$.
Then $\angle CAB = 3\theta$ and $\angle ACD = 180° - 4\theta > 0°$. This class consists of all triangles in which two of the angles are in the ratio 1:3, where the smaller angle θ satisfies $0° < \theta < 45°$.

Case 3. $\angle ACD = \angle CAD$.
Then their common value is $90° - \theta$ so that $\angle CAB = 90°$. This class consists of all right triangles.

6. Dissect a square of suitable size into 6 similar rectangles so that there are 1 large rectangle, 3 medium rectangles and 2 small rectangles, each with integral dimensions.

Solution:
We start with a rectangle R and divide it into 3 rectangles. R_1 is the right half of R, R_2 is the bottom half of $R - R_1$ and R_3 is $R - R_1 - R_2$. The dimensions of these rectangles will be adjusted later. Divide R_2 into 3 congruent rectangles using horizontal lines and R_3 into two congruent rectangles using vertical lines. Let the horizontal and vertical dimensions of each rectangle in R_i be x_i and y_i respectively, as shown in Figure A.8.

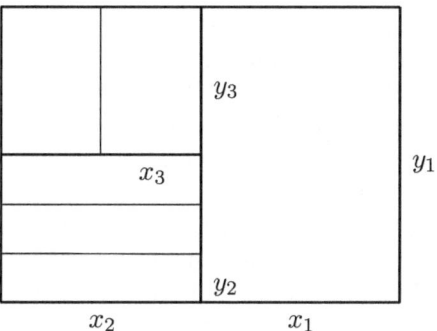

Figure A.8

Next, we make all the pieces similar to one another. This is done by setting $y_i = x_i t$ for all i, where t is some positive number. We choose $x_3 = 1$, so that $y_3 = t$. Now $x_2 = 2x_3 = 2$ and $y_2 = 2t$. Similarly, $y_1 = 3y_2 + y_3 = 7t$ and $x_1 = 7$. Finally, we want to choose t so that R is a square. From $7t = 2 + 7$, we have $t = \frac{9}{7}$. To obtain integral dimensions, we magnify the diagram by a factor of 7. Figure A.9 shows the desired dissection of a 63×63 square.

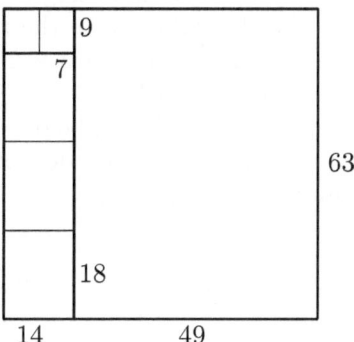

Figure A.9

7. There is a 3×10 hole of depth $\frac{1}{2}$ on a wall. In how many ways can it be filled by 15 rectangular bricks each of thickness $\frac{1}{2}$ and cross-section 1×2 or 2×1?

Solution:
Let a_n be the number of different ways of filling in a $3 \times 2n$ hole. Then $g_0 = 1$ and $g_1 = 3$. Some of the fillings of the $3 \times 2n$ hole can be divided by a vertical line into two parts without splitting any dominoes. Such a line is called a *fault line*. Those fillings without fault lines are said to be *fault-free*. Let b_n be the number of fault-free fillings of the $3 \times 2n$ hole. We have $b_1 = 3$. For all $n \geq 2$, a fault-free filling cannot start with three 1×2 bricks. It must start off as shown in Figure A.10, and continue by adding horizontal dominoes except for a final vertical one. It follows that $b_n = 2$ for all $n \geq 2$

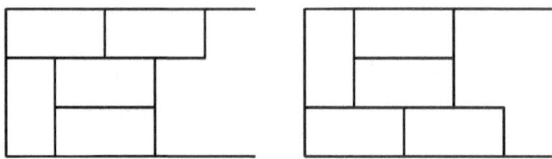

Figure A.10

The a_n fillings can be classified according to where the first fault line is. This is taken to be the right end of the hole if the filling is fault-free. Then the hole is divided into a $3 \times 2k$ part on the left and a $3 \times 2(n-k)$ part on the right where $1 \leq k \leq n$. Since the first part is filled without any fault lines, it can be done in b_k ways. The second part can be filled in a_{n-k} ways as we do not care whether there are any more fault lines. Hence $a_n = b_1 a_{n-1} + b_2 a_{n-2} + \cdots + b_n a_0$. We have

$$
\begin{array}{rlllll}
a_n & = & 3a_{n-1} & +2a_{n-2} & +2a_{n-3} & +\cdots & +2a_0, \\
a_{n-1} & = & & 3a_{n-2} & +2a_{n-3} & +\cdots & +2a_0; \\
\hline
a_n - a_{n-1} & = & 3a_{n-1} & -a_{n-2}. &&&
\end{array}
$$

This simplifies to $a_n = 4a_{n-1} - a_{n-2}$. Iteration yields $a_2 = 11$, $a_3 = 41$, $a_4 = 153$ and $a_5 = 571$.

8. A V-tromino is a 2×2 square with one of the 4 1×1 square missing. Prove that if any of the 49 1×1 square is removed from a 7×7 board, the remaining part can be covered by 16 copies of the V-tromino.

Solution:

By symmetry, the square removed must come from the shaded 2×2 subboard of one of the three boards shown in Figures A.12, A.13 and A.14. They also show that the remaining part of the board may be covered by V-trominoes. Whichever square is removed from the 2×2 subboard, the remaining three squares form a copy of the V-tromino and can be covered.

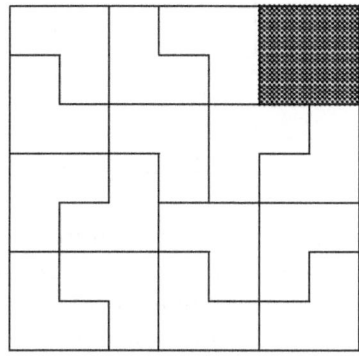

Figure A.11

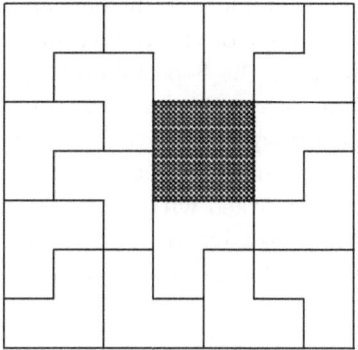

Figure A.12

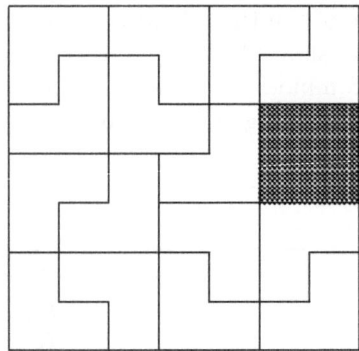

Figure A.13

9. An L-tetromino is a 2×3 rectangle missing one corner square and an adjacent non-corner square. Determine the minimum number n of colors for which there exists an n-color infinite infinite chessboard such that wherever an L-tetromino is placed, the four squares it covers all have different colors.

Solution:

Consider the region in Figure A.14, with 12 unit squares. Every two of the 4 central squares may be covered by a suitable placement of the L-Tetromino. Thus 4 colors, 1, 2, 3 and 4, are needed there. Moreover, any of these squares and any of the peripheral squares may be covered by a suitable placement of the L-Tetromio. Thus these 4 colors may not be used again for the 8 peripheral squares. If only 7 colors are available, then one of the 3 additional colors, say 5, must be used on at least 3 peripheral squares. Paint any peripheral square in color 5. We may assume by symmetry that it is the one shown in Figure A.14. Then the 4 squares marked with crosses cannot be painted in color 5. Now 2 of the 3 blank squares must be in color 5, but any 2 of them may be covered by a suitable placement of the L-Tetromino. This shows that $n \geq 8$.

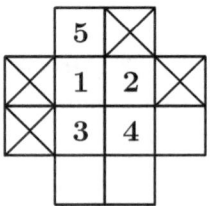

Figure A.14

Figure A.15 shows an 8-color infinite grid which establishes that $n = 8$ for the L-Tetromino.

1	2	5	6	1	2
3	4	7	8	3	4
5	6	1	2	5	6
7	8	3	4	7	8
1	2	5	6	1	2
3	4	7	8	3	4

Figure A.15

10. Consider a block of three stamps A, B and C in a row. It has six connected subblocks, namely, A, B, C, AB, BC and ABC. If the values of A, B and C are 1, 3 and 2 respectively, then the values of the six blocks are six consecutive integers, starting with 1. Such a block of stamps is said to be perfect. Find a perfect blocks consisting of five stamps, not necessarily in a row. Two adjacent stamps must meet along an entire edge.

Solution:
Figure A.16 shows a perfect block of five stamps. It is unique apart from the trivial interchange of 1 and 2.

3	2	
1	8	7

Figure A.16

11. A G4G7 number is defined to be a positive integer with at least 4 digits such that each digit is at least 7, that is, each is 7, 8 or 9. Find infinitely many pairs of G4G7 numbers whose products are also G4G7 numbers.

Solution:
We claim that $\overline{8\cdots88887} \times \overline{9\cdots98877} = \overline{8\cdots878887\cdots79899}$ for all non-negative integer n, where n is the number of digits under each bar. For $n = 0$, we can verify directly that $8887 \times 8877 = 78889899$.

Suppose the result holds for some $n \geq 0$. In the next case, we have

$$
\begin{aligned}
&\overline{88\cdots8}8887 \times \overline{99\cdots9}8887 \\
= \; &\overline{80\cdots0}0000 \times \overline{90\cdots0}0000 + \overline{80\cdots0}0000 \times \overline{9\cdots9}8877 \\
&+\overline{90\cdots0}0000 \times \overline{8\cdots8}8887 + \overline{8\cdots8}8887 \times \overline{9\cdots9}8877 \\
= \; &\overline{720\cdots0}00000\overline{\cdots0}0000 + \overline{79\cdots9}10160\overline{\cdots0}0000 \\
&+\overline{70\cdots9}99830\overline{\cdots0}0000 + \overline{8\cdots8}78887\overline{\cdots7}9899 \\
= \; &\overline{88\cdots8}788877\overline{\cdots7}9899.
\end{aligned}
$$

This completes the induction argument.

12. A and B are 20 kilometers apart, and are moving towards each other, both at 10 kilometers per hour. At the same time, a bee starts off from A and flies towards B. When it reaches B, it immediately turns around and flies back towards A. It goes back and forth between A and B until they meet. When flying from A towards B, the bee's speed is 18 kilometers per hour. When flying from B to A, it is 12 kilometer per hour. How far has the bee flown when A and B meet?

Solution:
Since A and B will meet after 1 hour, the bee has flown a total of 1 hour. Denote the direction from A to B as positive, so that the direction from B to A is negative. Let t be the total amount of time the bees flies in the positive direction, so that the total amount of time it flies in the negative direction is $1 - t$. So the total net distance (positive minus negative) flown by the bee is $18t - 12(1-t)$. In between two consecutive times the bee is with A, the net distance it has flown is exactly the distance advanced by A, regardless of where it meets B. Since the total distance advanced by A in 1 hour is 10 kilometers, we have $18t - 12(1 - t) = 10$ so that $t = \frac{11}{15}$.

13. A girls' school had 18 students in the graduating class. They invited a number of students from a boys' school nearby to their graduation dance. Each of the 18 girls danced with at most 3 boys. Each boy danced with at least 3 girls. What was the minimum number of girls such that each boy had danced with at least one of them?

Solution:
We first construct an example to show that 9 girls are necessary. In this graduation dance, there are 18 boys and 18 girls. Each girl dances with exactly 3 boys and each boy dances with exactly 3 girls. The girls are numbered from 1 to 18, and a boy is identified by the trio of girls with whom he dances.

(1,2,3)	(7,8,9)	(13,14,15)
(1,2,4)	(7,8,10)	(13,14,16)
(1,5,6)	(7,11,12)	(13,17,18)
(2,5,6)	(8,11,12)	(14,17,18)
(3,4,5)	(9,10,11)	(15,16,17)
(3,4,6)	(9,10,12)	(15,16,18)

Note that the second column is obtained from the first column by adding 6, and the third column from the second column also by adding 6. Among the girls 1 to 6, at least 3 of them must be chosen. This is because every two of these girls have at least one common dance partner, so that between them, they can dance with at most 5 of the 6 boys in the first column. Similarly, at least 3 of the girls 7 to 12 and at least 3 of the girls 13 to 18 must be chosen, yielding a total of 9.

Next, we prove that we never need to choose more than 9 girls. We apply the following protocol. Initially, we have 18 girls and x boys. In the first stage, we choose a girl who has danced with exactly 3 boys, and take her aside along with her dance partners. Then we choose an girl who has danced with exactly 3 of the remaining boys, and continue until no such girl exists any more. Let the number of girls chosen at this stage be a, and the let the number of boys left be y. We have $x - y = 3a$. In the second stage, we choose a girl who has danced with exactly 2 of the remaining boys, and take her aside along with her dance partners. We continue until no such girl exists any more. Let the number of girls chosen at this stage be b, and the let the number of boys left be z. We have $y - z = 2b$. In the third stage, we choose a girl who has danced with exactly 1 of the remaining boys, and take her aside along with her dance partner. We continue until no such girl exists any more. Let the number of girls chosen at this stage be c. We have $z = c$. We now count the total number of dances overall. Since each of the x boys dances with at least 3 girls, it is at least $3x$. Since each of the 18 girls dances with at most 3 boys, it is at most $3 \times 18 = 54$. It follows that $x \leq 18$. We next count the total number of dances not involving the girls chosen in the first stage. Since each of the y boys dance with at least 3 girls, it is at least $3y$. Since each of the $18 - a$ girls dances with at most 2 boys, it is at most $2(18 - a) = 36 - 2a$. It follows that $2a + 3y \leq 36$. Finally, we count the total number of dances not involving the girls chosen in the first two stages. Since each of the z boys dance with at least 3 girls, it is at least $3z$. Since each of the $18 - a - b$ girls dances with at most 1 boy, it is at most $18 - a - b$. It follows that $a + b + 3z \leq 18$. Now $3a + 2b + c = (x - y) + (y - z) + z = x \leq 18$ and $2a + 6b + 3c = 2a + 3(y - z) + 3z = 2a + 3y \leq 36$.

Hence

$$
\begin{array}{rcccccccc}
8 & (& 3a & + & 2b & + & c &) & \le & 8 & \times & 18 \\
2 & (& 2a & + & 6b & + & 3c &) & \le & 2 & \times & 36 \\
7 & (& a & + & b & + & 3c &) & \le & 7 & \times & 18 \\
\hline
35 & (& a & + & b & + & c &) & \le & 19 & \times & 18
\end{array}
$$

This yields $a+b+c \le \frac{342}{35} < 10$. Since a, b and c are positive integers, $a + b + c \le 9$.

14. Seated in a circle are 8 wizards. A positive integer not exceeding 100 is pasted onto the forehead of each. The numbers need not be distinct. A wizard can see the numbers of the other 7, but not his own. After ten minutes, a bell rings. Simultaneously, each wizard puts up either his left hand or his right hand. After another ten minutes, the bell rings again. Simultaneously, each wizard declares the number on his forehead. Find a strategy on which the wizards can agree beforehand, which can guarantee that each of them will make the correct declaration?

Solution:
A positive integer not exceeding 100 has 7 binary digits, if we include leading 0s, and there are 7 numbers visible to each wizard. Thus he has enough information to construct a 7×7 table, the rows being the binary representations of those 7 numbers in clockwise order after himself. Let the wizards be numbered from 1 to 8 in clockwise order. As an illustration, suppose wizard #1 sees the numbers 6, 53, 53, 34, 37, 46 and 73 on the foreheads of the others in clockwise order after himself. In the first ten minutes, he obtains the following table.

$$
\begin{array}{ccccccc}
\mathbf{0} & 0 & 0 & 0 & 1 & 1 & 0 \\
0 & \mathbf{1} & 1 & 0 & 1 & 0 & 1 \\
0 & 1 & \mathbf{1} & 0 & 1 & 0 & 1 \\
0 & 1 & 0 & \mathbf{0} & 0 & 1 & 0 \\
0 & 1 & 0 & 0 & \mathbf{1} & 0 & 1 \\
0 & 1 & 0 & 1 & 1 & \mathbf{1} & 0 \\
1 & 0 & 0 & 1 & 0 & 0 & \mathbf{1}
\end{array}
$$

The wizards agree beforehand the following rule. If the sum of the 7 binary digits in a wizard's diagonal is odd, that wizard puts up his left hand. If the sum is even, he puts up his right hand. In our example, the sum of the 7 digits on the diagonal of wizard #1 is 0+1+1+0+1+1+1, which is odd. Hence he will put up his left hand. In the second ten minutes, wizard #1 can figure out the binary representation of his own number. Suppose it is ABCDEFG. Consider for instance wizard #6. His table is the following.

$$
\begin{array}{ccccccc}
\mathbf{0} & 1 & 0 & 1 & 1 & 1 & 0 \\
1 & \mathbf{0} & 0 & 1 & 0 & 0 & 1 \\
A & B & \mathbf{C} & D & E & F & G \\
0 & 0 & 0 & \mathbf{0} & 1 & 1 & 0 \\
0 & 1 & 1 & 0 & \mathbf{1} & 0 & 1 \\
0 & 1 & 1 & 0 & 1 & \mathbf{0} & 1 \\
0 & 1 & 0 & 0 & 0 & 1 & \mathbf{0}
\end{array}
$$

Now wizard #1 knows 6 of the 7 digits on the diagonal of the table of wizard #6, the sum of which is 0+0+0+1+0+0, an odd number. So if wizard #6 puts up his left hand, wizard #1 will know that C=0. If wizard #6 puts up his right hand, wizard #1 will know that C=1. In the same way, wizard #1 can determine the other digits of the binary representation of his own number, because each lies on the diagonal of the table of a different wizard. Specifically, A lies on the diagonal of the table of wizard #8, B on #7, D on #5, E on #4, F on #3 and G on #2. Thus wizard #1 can confidently declare the correct value of his number. What works for wizard #1 works for each of the others, as the situation has cyclic symmetry. Therefore, the simple convention of "odd-left even-right" does the trick.

15. A tournament is a complete directed graph, meaning that between any two vertices, there is exactly one arc pointing from one of them to the other. Three vertices X, Y and Z are said to form a 3-cycle if the arcs go from X to Y to Z and back to X, or X to Z to Y and back to X. In a tournament with n players, let the number of arcs going out of the i-th vertex be w_i for $1 \le i \le n$. Prove that the number of 3-cycles is given by

$$
\frac{n(n-1)(2n-1)}{12} - \frac{1}{2}\sum_{i=1}^{n} w_i^2.
$$

Solution:
A *broken arrow* in a directed graph is defined as the union of two arcs with a common vertex, which is called the *pivot*, such that one arc goes into the pivot and the other goes out of it. Now the i-th vertex has w_i outgoing arcs, so that it has $n - 1 - w_i$ incoming arcs. It is the pivot of exactly $w_i(n - 1 - w_i)$ broken arrows. Since $\sum_{i=1}^{n} w_i = \binom{n}{2}$, the total number of broken arrows is

$$
\sum_{i=1}^{n} w_i(n - 1 - w_i) = (n-1)\binom{n}{2} - \sum_{i=1}^{n} w_i^2.
$$

Of the $\binom{n}{3}$ triples of vertices, let there be λ 3-cycles. Each of them contains 3 broken arrows while each of the other triples contains only 1. It follows that the total number of broken arrows is also given by $3\lambda + \binom{n}{3} - \lambda = 2\lambda + \binom{n}{3}$. Hence

$$(n-1)\binom{n}{2} - \sum_{i=1}^{n} w_i^2 = 2\lambda + \binom{n}{3}$$

so that

$$
\begin{aligned}
\lambda &= \frac{1}{2}\left[(n-1)\binom{n}{2} - \binom{n}{3} - \sum_{i=1}^{n} w_i^2 \right] \\
&= \frac{n(n-1)^2}{4} - \frac{n(n-1)(n-2)}{12} - \frac{1}{2}\sum_{i=1}^{n} w_i^2 \\
&= \frac{n(n-1)}{12}[3(n-1) - (n-2)] - \frac{1}{2}\sum_{i=1}^{n} w_i^2 \\
&= \frac{n(n-1)(2n-1)}{12} - \frac{1}{2}\sum_{i=1}^{n} w_i^2.
\end{aligned}
$$

16. How many convex polyhedra are there with five vertices?

Solution:
If a polyhedron has exactly 5 vertices, the set of degrees of its vertices may be {3,3,3,3,3}, {3,3,3,4,4}, {3,4,4,4,4}, {3,3,3,3,4} or {4,4,4,4,4}. However, the first three sets are not feasible, since the sum of all the degrees must be an even number, equal to twice the number of edges. The fourth is a square pyramid, shown on the left side of Figure A.17. The fifth is a double triangular pyramid, shown on the right side of Figure A.17. The last set is not the skeleton of a polyhedron because it is the non-planar graph K_5.

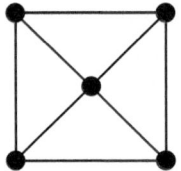 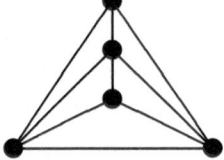

Figure A.17

Appendix B: Solution to Exercises

Chapter 1:

1. Figure 1.2 shows a five-pointed star consisting of five lines intersecting one another in ten points. At each point, we place a message digit, and then add a parity-check digit for each line. Thus the digits A to K are used to convey the message. The digits L to Q are chosen so that the number of 1s on each line is even.

   As an illustration, suppose the received message is as shown in the diagram. The parity-check fails only on the horizontal line. It follows that a single transmission error occurs at the digit N. This code was published in [8].

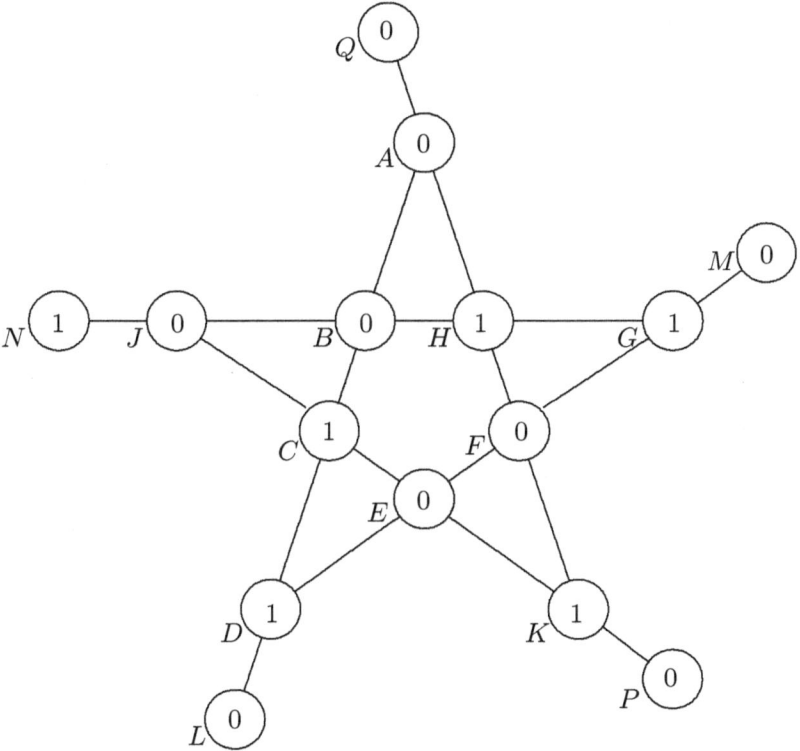

Figure 1.2

2. There are three possible inputs for each prisoner, seeing two white hats, seeing two black hats and seeing one hat of each color. There are also three possible actions for each prisoner, pass, guess white or guess black. It is reasonable to expect a one-to-one correspondence between

© Springer International Publishing AG 2018
A. Liu, *S.M.A.R.T. Circle Projects*, Springer Texts
in Education, DOI 10.1007/978-3-319-56811-9

inputs and actions. By symmetry, the prisoner should pass if he sees one hat of each color.

Suppose he sees two hats of the same color. If he guesses the same color, he will only be right if all three hats are of the same color, which happens only $\frac{1}{4}$ of the time. Hence he should guess the opposite color. Using this scheme, if all three hats are of the same color, all three prisoners will guess and all three will be wrong. However, if the three hats are not all of the same color, only the prisoner wearing the hat of a color different from the other two will guess, and he will be right. Since the three hats are not all of the same color $\frac{3}{4}$ of the time, this is their probability of going free. The astute reader may recognize this as the Hamming code for a three-digit transmitter.

3. The total number of scenarios we have to distinguish is

$$\binom{15}{0} + \binom{15}{1} + \binom{15}{2} + \binom{15}{3} = 576.$$

Since $2^9 = 512 < 576$, 9 check digits will not be sufficient.

Chapter 2:

1. Suppose there is a polyhedron with $E = 7$. Cutting the edges at their midpoints, each of the V vertices is attached to at least 3 half-edges Hence $3V \leq 2E = 14$ so that $V \leq 4$. Cutting the edges along their midlines, each of the F face is bounded by at least 3 half-edges. Hence $3F \leq 2E = 14$, so that $F \leq 4$. By Euler's Formula, $2 = V + F - E \leq 4 + 4 - 7 = 1$, which is a contradiction.

2. (a) First form a doubled triangle with both hands of one student and one hand of the other student. Then the other student uses the other hand to pull three segments of the string together, as shown in Figure 2.30 on the left. Then we have the skeleton of a tetrahedron. Actually, it is sufficient to pull only two of those three segments.

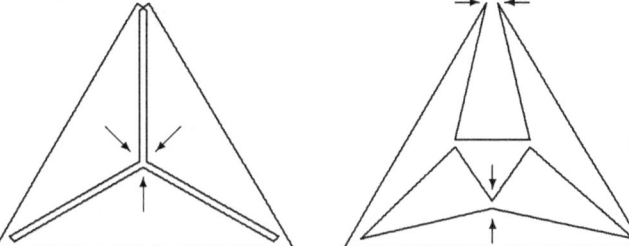

Figure 2.30

(b) First form a doubled quadrilateral with both hands of two students. Then the third student uses each hand to pull two segments of the string together, as shown in Figure 2.30 on the right. Then we have the skeleton of a standard octahedron.

(c) First form a doubled quadrilateral with both hands of two students,. Then the other two students use each hand to pull two segments of the string together, as shown in Figure 2.31. Then we have the skeleton of a cuboid.

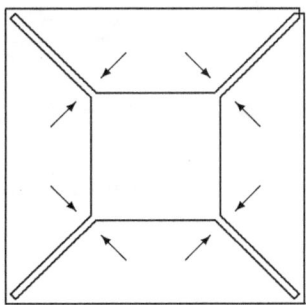

Figure 2.31

3. (a) Figure 2.32 shows one orientation of the snub cube.

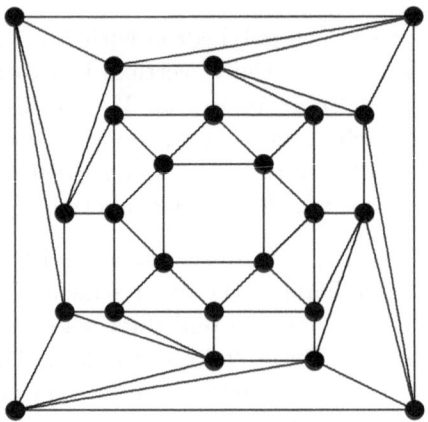

Figure 2.32

(b) Figure 2.33 shows one orientation of the snub dodecahedron.

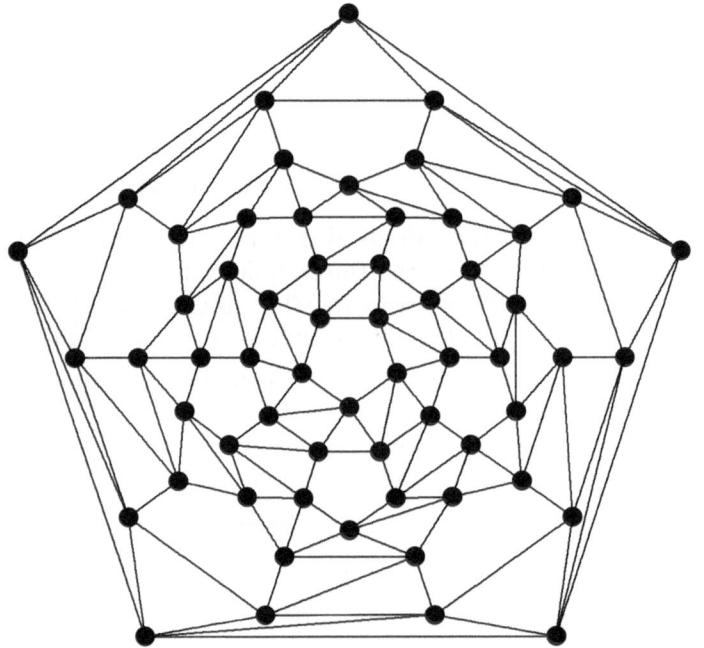

Figure 2.33

Chapter 3:

1. (a) Figure 3.22 shows that the L-tetromino has the desired property.

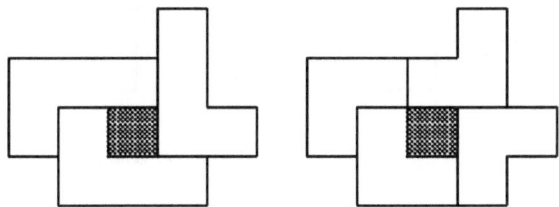

Figure 3.22

   (b) The I- and O-tetraminoes may be dismissed immediately as each copy can only touch one side of the hole. Figure 3.23 on the left shows where the hole must be relative to the S- and T-tetrominoes, each with five blank squares to be covered. The S-tetrominoes can now be dismissed since each copy can cover only two of those squares. Finally, Figure 3.23 on the right shows the only two possible positions of the other two copies of the T-tetromino which can cover the five squares. However, neither figure can be covered by the V-Tromino. Hence the L-tetromino is the only one with the desired property.

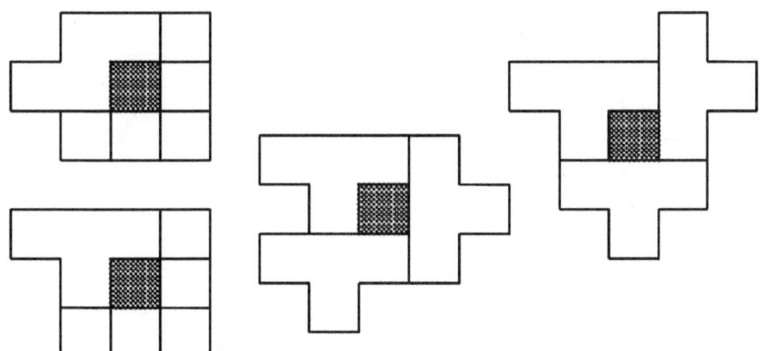

Figure 3.23

2. The domino cannot be used in the equation $2x = 2y$, and we have seen that there is a solution with x and t being the two trominoes. The domino cannot be used in the equation $2x = x + y$ either, but this time, it is not hard to verify that there are no solutions. The equation $2w = x + y$ cannot have a solution regardless of whether the domino is w, x or y. In the equation $w + x = w + y$, the domino must be w, and there are three solutions, as shown in Figure 3.24. Finally, we simply do not have enough polyominoes for the equation $x + y = u + v$.

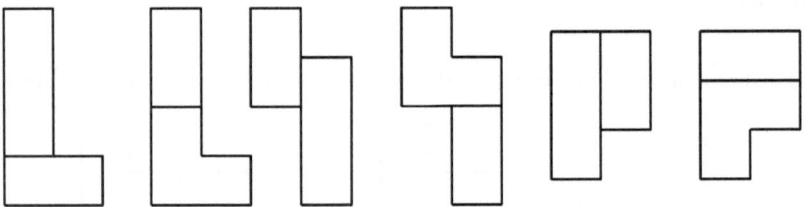

Figure 3.24

3. (a) Figure 3.25 shows that the triamond is compatible with all three tetriamonds.

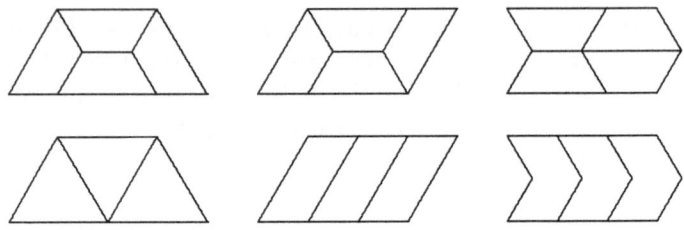

Figure 3.25

(b) Figure 3.26 shows that the triamond is compatible with all four pentiamonds.

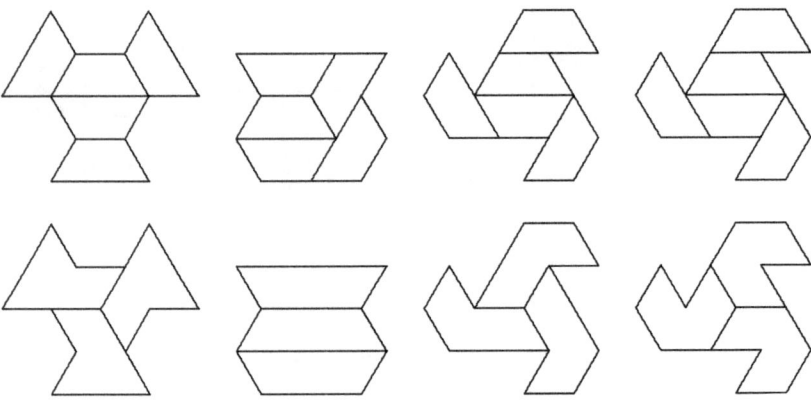

Figure 3.26

Chapter 4:

1. Figure 4.19(a) shows a closed tour for Timmy, where the squares are numbered in the order in which they are visited. If we shift the second column to the far right and the third column to the far left, we obtain Figure 4.19(b). If we then shift the second row to the bottom and the third row to the top, we obtain Figure 4.19(c), which is a desired closed tour for the King's Rook.

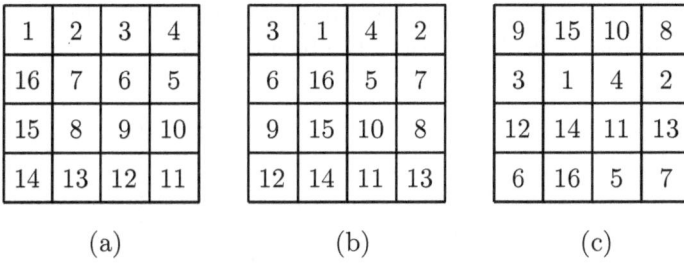

| (a) | (b) | (c) |

Figure 4.19

2. White's winning strategy is summarized in the following chart.

Moves	(1)	(2)	(3)	(4)	(5)	(6)	(7)	(8)	(9)	(10)
White	f8	f6	h6	h7	g6	g7	h7	g8	g6	h7
Red	h7	g8	f7	f8	e7	e6	f6	e7	f8	e8
Notes	(a)			(b)						

Notes:
(a) If (1) ... g6, then (2) h8 f7, and continue from (4).
(b) If (4) ... e6, continue as before. If (4) ... f6, continue from (8).

3. We suppose that there is a re-entrant knight tour on the $4 \times n$ board. If we put $2n$ Knights on the board so that none of them attacks another, they must occupy alternate positions along the re-entrant tour. There are only two such sets of positions, so that there can only be two ways of placing the non-attacking Knights. However, we can put the $2n$ Knights all on white squares, all on black squares, or all on the top and bottom rows. This means that there cannot be a re-entrant tour on the $4 \times n$ board.

Chapter 5:

1. The result is trivially true for $n = 0$. Suppose it holds for some $n \geq 0$. Consider a $2^{n+1} \times 2^{n+1}$ chessboard. Divide it into four equal quadrants. Without loss of generality, we may assume that the missing square is in the north-east quadrant. By the induction hypothesis, the rest of this quadrant may be covered by copies of the given shape. We now place a copy of the given shape as shown in Figure 5.9. Then each of these quadrant is missing one square. By the induction hypothsis again, each of them may be covered by the copies of the given shape. By mathematical induction, the result holds for all $n \geq 1$.

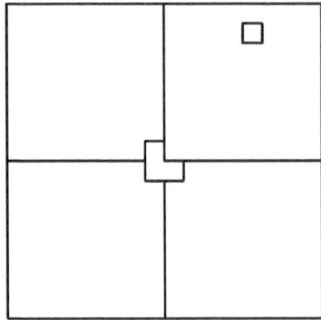

Figure 5.9

2. (a) We have $a_{13+1} = 39 = 3(13)$ and $d_{13} = 13$ is prime. The smallest prime divisor of $2(13) - 1 = 25$ is 5, and the next such value of n occurs at $13 + \frac{5-1}{2} = 15$.

    (b) We have $a_{17+1} = 51 = 3(17)$ and $d_{13} = 17$ is prime. The smallest prime divisor of $2(17) - 1 = 33$ is 3, and the next such value of n occurs at $17 + \frac{3-1}{2} = 18$.

    (c) We have $a_{19+1} = 57 = 3(19)$ and $d_{13} = 19$ is prime. The smallest prime divisor of $2(19) - 1 = 37$ is 37 itself, and the next such value of n occurs at $19 + \frac{37-1}{2} = 37$.

    (d) We have $a_{23+1} = 69 = 3(23)$ and $d_{23} = 23$ is prime. The smallest prime divisor of $2(23) - 1 = 45$ is 3, and the next such value of n occurs at $23 + \frac{3-1}{2} = 24$. The sequences $\{a_n\}$ and $\{d_n\}$ coincide with those generated by $a_1 = 3$ from this point on.

3. (a) The chest may be opened in two steps.

    - Touch a pair of drums. If both monkeys are right side up, hit no drums. If one of them is right side up, hit the other drum. If both are upside down, hit both drums. The end result is that both monkeys are right side up.

- Suppose the chest is not open. Touch a pair of drums. The monkeys cannot both be upside down. If both are right side up, hit both drums. If one of them is right side up, hit the other drum. The chest will open.

(b) The chest may be opened in five steps.

- Touch a pair of opposite drums and make both monkeys right side up.

- Suppose the chest is not open. Touch a pair of adjacent drums and make both monkeys right side up.

- Suppose the chest is not open. Touch a pair of opposite drums. The monkeys cannot both be upside down. If one of them is upside down, hit that drum and the chest will open. If both are right side up, hit either drum.

- Touch a pair of adjacent drums. If both monkeys are of the same posture, hit both drums and the chest will open. If they are of opposite postures, hit both drums.

- Touch a pair of opposite drums. The monkeys will have the same posture. Hitting both drums will open the chest.

Chapter 6:

1. We append terms of $k + 1$ if necessary so that all labels are of length $m - 1$. Write down a number of 0s equal to the number of 1s in the label. Insert a 1 after this block. Then write down a number of 0s equal to the number of 2s, followed by another 1, and so on. Note that each binary sequence consists of $k + 1$ 1s and $m - 1$ 0s, starts with a 0 and ends with a 1. The k 1s in between the first and the last terms can be placed in $\binom{k+m-2}{k}$ ways.

2. A simpler Diamond Formula is $South = (West + East + 1) - North$ We have .

$$
\begin{aligned}
&(West + East + 1) - North \\
=\ &((m(n + 1) + 1) + ((m + 1)n) + 1) + 1) - (mn + 1) \\
=\ &mn + m + 1 + mn + n + 1 + 1 - mn - 1 \\
=\ &mn + m + n + 2 \\
=\ &(m + 1)(n + 1) + 1 \\
=\ &South.
\end{aligned}
$$

3. Suppose an order 6 TOAD exists. We partition its twenty-one numbers into seven triples, identified by different letters, as shown in Figure 6.17. Each triple contains an even number of odd numbers. Hence the TOAD contains an even number of odd numbers. However, there are eleven odd numbers from 1 to 21, and we have a contradiction.

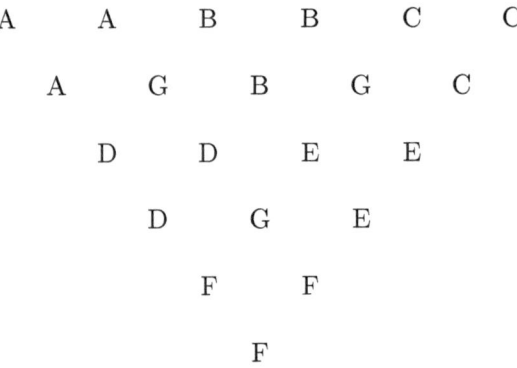

$$
\begin{array}{ccccccc}
A & & A & & B & & B & & C & & C \\
& A & & G & & B & & G & & C & \\
& & D & & D & & E & & E & & \\
& & & D & & G & & E & & & \\
& & & & F & & F & & & & \\
& & & & & F & & & & &
\end{array}
$$

Figure 6.17

Chapter 7:

1. Express the first 128 non-negative integers in base 2 with leading 0s so that each has 7 binary digits. The condition of the problem is that no two 1s are adjacent. Let the number of such integers be a_n where n is the maximum number of binary digits. Then $a_1 = 2$, $a_2 = 3$ and $a_3 = 5$. The integers with at most 3 digits are 000, 001, 010, 100 and 101. For $n \geq 4$, if the last digit is 0, the number of such integers is a_{n-1}. If the last digit is 1, then the second last digit must be 0, and the number of such integers is a_{n-2}. It follows that $a_n = a_{n-1} + a_{n-2}$, so that $a_4 = 8$, $a_5 = 13$, $a_6 = 21$ and $a_7 = 34$. The largest 7-digit number $1010101_2 = 85$ is under 100. Taking away 0, we have 33 such positive integers.

2. Since $336 = 7 \times 48$ and $7 = 3 + 4$, $336 + 3 + 3 + \cdots + 3 + 4 + 4 + \cdots + 4$, where there are 48 copies of 3 and 48 copies of 4. Adding 0, 1, 2, ..., 47 to the 4s respectively, the last term is 51. Subtracting 47, 46, 45, ..., 0 from the 3s, the first term is -44. After cancellation of the negative terms, we have $336 = 45 + 46 + 47 + 48 + 49 + 50 + 51$.

3. (a) Clearly, two beetles lined up directly in front of the target square can serve as the escape team. A team of size one is insufficient, because the maximum value of the lone beetle is x, and we have $x < x + x^2 = 1$.

   (b) Four beetles positioned as shown in Figure 7.8 can serve as the escape team. After the first 2 moves, we are at the scenario in (a). A team of size three is insufficient, because the maximum total value of the beetles is $x^2 + 2x^3 < 2x^2 + x^3 = x + x^2 = 1$.

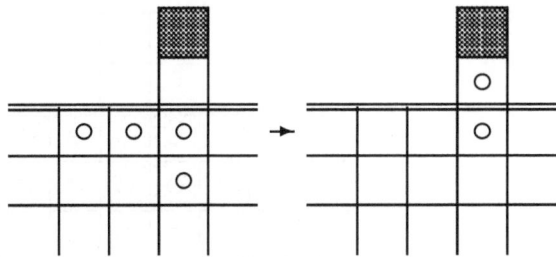

Figure 7.8

   (c) Eight beetles positioned as shown in Figure 7.9 can serve as the escape team. The first 3 moves are exactly as in (b). A team of size seven is insufficient, because the maximum total value of the beetles is

$$x^3 + 3x^4 + 3x^5 < x^3 + 4x^4 + 2x^5 = 3x^3 + 2x^4 = 2x^2 + x^3 = 1.$$

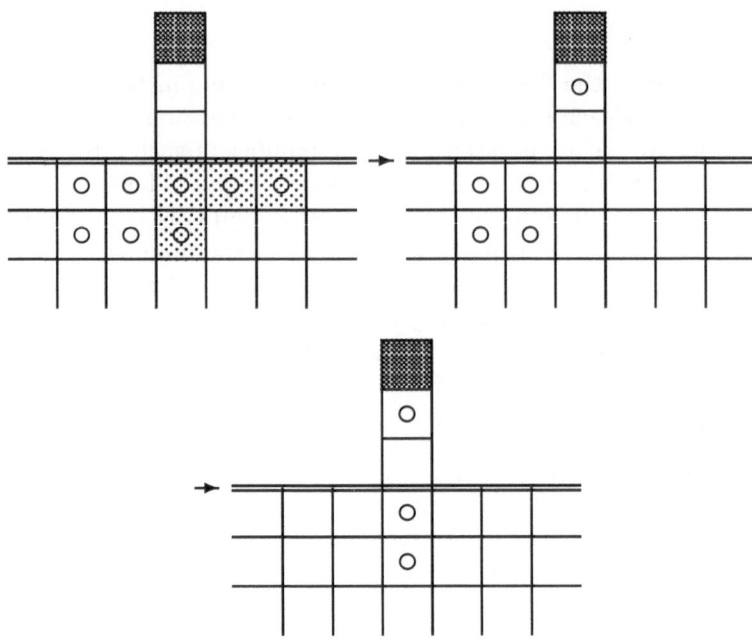

Figure 7.9

(d) Twenty beetles positioned as shown in Figure 7.10 can serve as the escape team. The first 7 moves are exactly as in (c). We also use the 3 moves in (b) three times.

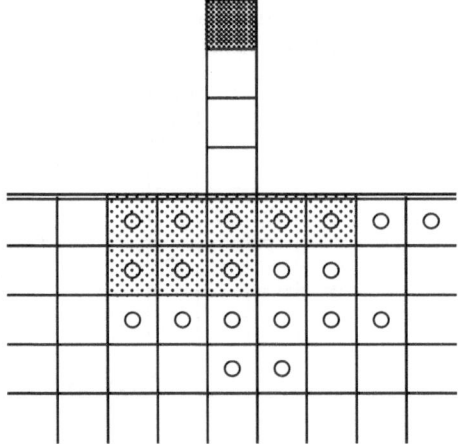

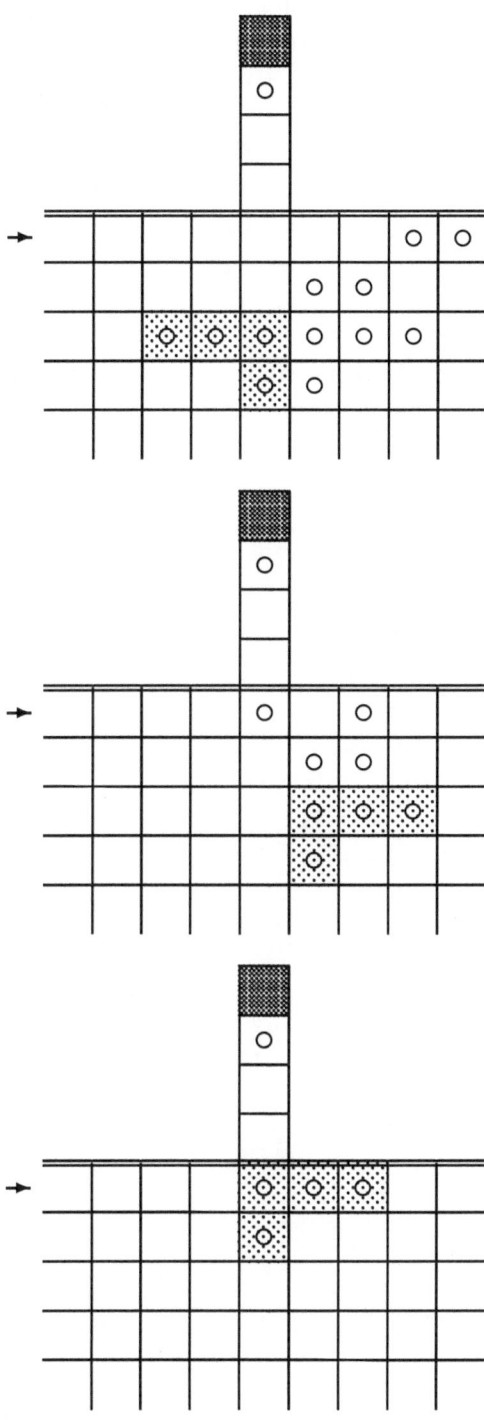

Figure 7.10

An escape team of size nineteen may just be sufficient, because the maximum total value of the beetles is

$$
\begin{aligned}
& x^4 + 3x^5 + 5x^6 + 7x^7 + 3x^8 \\
=\ & x^4 + 3x^5 + 8x^6 + 4x^7 \\
=\ & x^4 + 7x^5 + 4x^6 \\
=\ & 5x^4 + 3x^5 \\
=\ & 3x^3 + 2x^4 \\
=\ & 1.
\end{aligned}
$$

However, this possibility may be eliminated by a brute force examination of cases.

Chapter 8:

1. (a) By Postulate 1, there are at least two lions. Consider one of them. By Postulate 2, this lion has bitten at least three ponies. It follows clearly that there are at least two ponies.

   (b) Consider any pony P. We have proved that there is a lion that has not bitten P. By Postulate 2, this lion has bitten three different ponies Q, R and S. By Postulate 4, there is a lion that has bitten both P and Q, a lion that has bitten both P and R and a lion that has bitten both P and S. By Postulate 3, these are three different lions, and they have all bitten P.

   (c) Consider any pair of ponies. By Postulate 4, there is at least one lion has bitten both of them. Suppose there are two or more such lions. Consider a pair of them. Then it is not true that there is exactly one pony that both have bitten. This contradicts Postulate 3. Hence there is exactly one lion that has bitten both ponies.

   (d) Consider any pair of lions. By Postulate 3, there is exactly one pony that both have bitten. Hence it is certainly true that there is at least one pony that both have bitten.

2. Take a copy of the Fano plane and labeled the points A, B, C, D, E, F and G as shown in Figure 8.2, and define the lines as before. Each line represents a Space Pod carrying all the parts whose labels do not appear on that line. Let the other five parts be H, I, J, K and L, with each of Space Pods 8 and 9 carrying all of them. In addition, Some of them are carried by the first seven Space Pods, as shown in the list below.

    Space Pod 1: D, E, F, G, H and I.
    Space Pod 2: A, B, F, G, H and J.
    Space Pod 3: A, B, C, D, I and J.
    Space Pod 4: B, C, E, F and K.
    Space Pod 5: A, B, D, E and K.
    Space Pod 6: A, C, D, F and L.
    Space Pod 7: A, C, E, G and L.
    Space Pod 8: H, I, J, K and L.
    Space Pod 9: H, I, J, K and L.

   In order for the Space Dodecapus to get all of A, B, C, D, E, F and G, it must capture three of Space Pods 1 to 7. Then, in order for the Dodecapus to get all of H, I and J, it must capture two of Space Pods 1 to 3. In order for the Space Dodecapus to get both K and L, it must capture one of Space Pods 4 and 5, and one of Space Pods 6 and 7. However, it can only capture three Space Pods.

3. To prove that $f(7, 3k + 1) \leq 7k + 2$, observe that the total number of stops is $21k + 7$. If each floor is served by at least 3 elevators, then the number of floors is at most $7k + 2$. If some floor is served by at most 2 elevators, it can be linked to at most $6k$ other floors. Counting this floor, the building can have at most $6k + 1$ floors. That $f(7, 3k + 1) \geq 7k + 1$ follows from Observation 2 and $f(7, 3k) = 7k$.

Chapter 9:

1. Label the jewels as shown in Figure 9.14. The only way the crew can beat Captain Crook is by getting the 5 and either the 3 or the 4. So either W_2 or C must be 5. Suppose W_2 is 5. Captain takes N or E, whichever has higher value. If the other one is at most 2, the crew can get at most 5+2=7. The only way the crew can win is if {N,E}={3,4}, in which case there is nothing Captain Crook can do about it. Suppose C is 5. If W_1 is 1 or 2, Captain Crook takes N or E, whichever has higher value. The crew must take W_1 or loses C. Either way, Captain Crook wins. Finally, if W_1 is 3 or 4, the crew wins. If Captain Crook does not take W_1, the crew will. If he does, the crew takes W_1, N or E, whichever has highest value. The overall winning probability for Captain Crook is $\frac{3}{5} + \frac{1}{5} \times \frac{1}{12} + \frac{1}{5} \times \frac{1}{2} = \frac{43}{60}$.

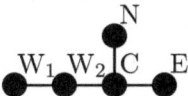

Figure 9.14

2. Let $m = qn+i$ where $0 \leq i < q$. Let the amount of gold in the t-th box be a_t kilograms and the amount of silver in the t-th box be b_t kilograms. We may assume that $a = a_1 \geq a_2 \geq \cdots \geq a_m$. Proceed as in the Proof of Lemma 2. Note that the difference between the amount of gold in the boxes in any two groups can never exceed a kilograms if we only redistribute the $((\ell-1)m+1)$-st, $((\ell-1)m+2)$-nd, ..., ℓm-th boxes among the groups in any way for any integer ℓ where $1 \leq \ell \leq \lambda$. Let the r-th group have the highest total amount of silver in the boxes in any group, and the s-th group the lowest. Let these total amounts be B_r and B_s kilograms respectively. Suppose that $B_r - B_s > b$. Then there must be some index ℓ such that $b_{\ell m+r} \geq b_{\ell m+s}$. Switch the $(\ell m + r)$-th and the $(\ell m + s)$-th boxes. As observed earlier, this does not mess up the gold situation, but the value of $B_r - B_s$, while still positive, has been reduced. Since this process can only be performed a finite number of time, we will eventually arrive at a distribution in which the difference between the maximum and minimum total amounts is as small as possible. Since further reduction will be possible if this difference is greater than b kilograms, we have the desired conclusion.

3. Captain Crook can get at least $\frac{3}{5}$ of the rum. He first subdivides 1 into $\frac{2}{5}$ and $\frac{3}{5}$. There are two cases.
   Case A. The crew subdivides $\frac{2}{5}$ into x and $\frac{2}{5} - x$, where $0 \leq x \leq \frac{1}{5}$. Captain Crook will subdivide $\frac{3}{5}$ into x and $\frac{3}{5} - x$. Now the four barrels are of sizes $x = x \leq \frac{2}{5} - x < \frac{3}{5} - x$.

No matter what the crew does, the size of the second largest barrel is at most $\frac{2}{5} - x$ and the size of the fourth largest barrel is at most x. Hence the crew gets at most $(\frac{2}{5} - x) + x = \frac{2}{5}$.

Case B. The crew subdivides $\frac{3}{5}$ into x and $\frac{3}{5} - x$, where $0 \le x \le \frac{3}{10}$. If $0 \le x \le \frac{1}{5}$, Captain Crook will subdivide $\frac{2}{5}$ into x and $\frac{2}{5} - x$, and the situation is exactly the same as in Case A. Hence we may assume that $\frac{1}{5} < x \le \frac{3}{10}$. Captain Crook will subdivide $\frac{3}{5} - x$ into $\frac{1}{5}$ and $\frac{2}{5} - x$. Now the four barrels are of sizes $\frac{2}{5} - x < \frac{1}{5} < x < \frac{2}{5}$. There are four subcases.

Subcase B1. The crew subdivides $\frac{2}{5}$ into y and $\frac{2}{5} - y$, where $0 \le y \le \frac{1}{5}$. Since $y + (\frac{2}{5} - y) = \frac{2}{5} = x + (\frac{2}{5} - x)$, the crew will get two barrels which add up to $\frac{2}{5}$.

Subcase B2. The crew subdivides x.

If $\frac{1}{5}$ remains the third largest barrel , Captain Crook get at least $\frac{2}{5} + \frac{1}{5} = \frac{3}{5}$. If it becomes the second largest barrel , the crew gets at most $\frac{1}{5} + \frac{1}{5} = \frac{2}{5}$.

Subcase B3. The crew subdivides $\frac{1}{5}$ into y and $\frac{1}{5} - y$, where $0 \le y \le \frac{1}{10}$. Since $\frac{2}{5} - x \ge y$, the second smallest barrel is at most $\frac{2}{5} - x$. Hence the crew gets at most $(\frac{2}{5} - x) + x = \frac{2}{5}$.

Subcase B4. The crew subdivides $\frac{2}{5} - x$.

Captain Crook get at least $\frac{2}{5} + \frac{1}{5} = \frac{3}{5}$.

We now prove that the crew can get at least $\frac{2}{5}$ of the rum. Captain Crook will first subdivide 1 into x and $1 - x$, where $0 \le x \le \frac{1}{2}$. There are three cases.

Case A. $\frac{2}{5} \le x \le \frac{1}{2}$.

The crew will subdivide $1 - x$ into x and $1 - 2x$. Now the three barrels are of sizes $1 - 2x < x = x$. If Captain Crook does not subdivide either x, neither will the crew. Then the crew will be sure of getting x plus another barrel, and $x \ge \frac{2}{5}$. If Captain Crook subdivides one of x, the crew will subdivide the other x in the same proportions. Then the crew will get two barrels which add up to $x \ge \frac{2}{5}$.

Case B. $\frac{1}{5} \le x < \frac{2}{5}$.

The crew will subdivide x into $x - \frac{1}{5}$ and $\frac{1}{5}$. Now the three barrels are of sizes $x - \frac{1}{5} < \frac{1}{5} < 1 - x$. If Captain Crook does not subdivide $1 - x$, the crew will subdivides this in halves. The second smallest barrel cannot be less than $\frac{1}{2}(x - \frac{1}{5})$, so the crew will get at least $\frac{1-x}{2} + \frac{1}{2}(x - \frac{1}{5}) = \frac{2}{5}$. Suppose Captain Crook subdivides $1 - x$ into y and $1 - x - y$, where $0 \le y \le \frac{1-x}{2}$. Then the crew will subdivide $1 - x - y$ into $\frac{2}{5} - y$ and $\frac{3}{5} - x$. Now $y + (\frac{2}{5} - y) = \frac{2}{5} = (x - \frac{1}{5}) + (\frac{3}{5} - x)$. Thus the crew will get two barrels which add up to $\frac{2}{5}$.

Case C. $0 \le x < \frac{1}{5}$.

The crew will subdivide $1 - x$ into $\frac{1}{5}$ and $\frac{4}{5} - x$. The situation is exactly the same as in Case B.

Chapter 10:

1. The modified circuit is shown in Figure 10.5.

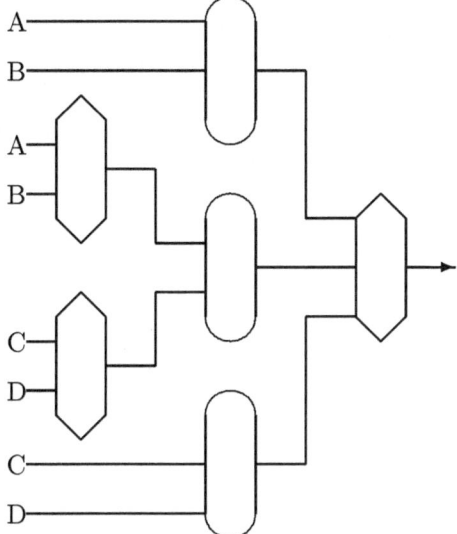

Figure 10.5

2. Divide the eight players into four pairs. Have each player play the other in the pair. That requires four hours. Now rank four of the players. Take two groups of two; call them X and Y. Have the better player of X play the better player of Y. The winner is the best of those two groups. The loser then plays the worse player of the other group. The winner of the second game is second best. At this point, if the loser of the first game loses the second, then you have enough information to rank all four players in X and Y. Otherwise, the worse players of both groups must play to determine who is third best. So, ranking four players consisting of two pairs of ranked players takes at most three hours. To get two ordered groups of four players takes at most six hours starting with ranked pairs.

   Now we have two ordered groups of four: ABCD and EFGH ordered from best to worst. To complete the ranking, We match the top player from each group. The winner takes the highest available place while the loser plays the next player in the other group. This continues until one group runs out of players. The longest this can take is if the bottom players of the two groups meet. This means that three players from each group have won matches, so that this match is the seventh one in this round. It follows that the total 4+6+7=17 hours.

3. In the first round, you go down N. All four campers go down E. When all reunite, if you have found the campsite, then everyone goes to N. If three or four campers agree on E, then it is E. If three or four campers agree that it isn't E, then you check W in the second round, and all campers rest. Suppose in the first round, two say it is down E and two say it isn't. Then you explore E again while sending one of those who said it isn't E down W. Now if you find it, then it is E. If not, then the camper who went down W must be telling the truth, because the liars were those who said it is down E.